GSM Networks: Protocols, Terminology, and Implementation

For a complete listing of the *Artech House Mobile Communications Library*, turn to the back of this book.

GSM Networks: Protocols, Terminology, and Implementation

Gunnar Heine

Artech House
Boston • London

Library of Congress Cataloging-in-Publication Data
Heine, Gunnar.
 [GSM—Signalisierung verstehen und praktisch anwenden. English]
 GSM networks : protocols, terminology, and implementation / Gunnar Heine
   p.   cm. — (Artech House mobile communications library)
 Translation of: GSM—Signalisierung verstehen und praktisch anwenden.
 Includes bibliographical references and index.
 ISBN 0-89006-471-7 (alk. paper)
 1. Global system for mobile communications.   I. Title.
TK5103.483.H4513   1998
621.3845'6—dc21   98-51784
                                                                              CIP

British Library Cataloguing in Publication Data
Heine, Gunnar
 GSM networks : protocols, terminology, and implementation—
 (Artech House mobile communications library)
 1. Global system for mobile communications
 I. Title
 621.3'8456

 ISBN 0-89006-471-7

Cover design by Lynda Fishbourne

© 1998 Franzis' Verlag GmbH
Translated from GSM - Signalisierung verstehen und praktisch anwenden
(Franzis' Verlag 1998)
English translation version:

© 1999 ARTECH HOUSE, INC.
685 Canton Street
Norwood, MA 02062

All rights reserved. Printed and bound in the United States of America. No part of this book may be reproduced or utilized in any form or by any means, electronic or mechanical, including photocopying, recording, or by any information storage and retrieval system, without permission in writing from the publisher.
 All terms mentioned in this book that are known to be trademarks or service marks have been appropriately capitalized. Artech House cannot attest to the accuracy of this information. Use of a term in this book should not be regarded as affecting the validity of any trademark or service mark.

International Standard Book Number: 0-89006-471-7
Library of Congress Catalog Card Number: 98-51784

10 9 8 7

Contents

1	**Introduction**	**1**
1.1	About This Book	1
1.2	Global System for Mobile Communication (GSM)	2
1.2.1	The System Architecture of GSM: A Network of Cells	3
1.2.2	An Overview on the GSM Subsystems	4
1.3	The Focus of This Book	7
1.4	Signaling	8
1.4.1	What is Signaling?	8
1.4.2	How is Signaling Performed?	8
1.4.3	What is Signaling Used For?	10
1.5	Representation of Messages	10
2	**The Mobile Station and the Subscriber Identity Module**	**13**
2.1	Subscriber Identity Module	13
2.1.1	The SIM as a Database	15
2.1.2	Advantage for the Subscriber	15
2.2	Mobile Station	17
2.2.1	Types of Mobile Stations	17

v

2.2.2	Functionality	17
2.2.3	Mobile Stations as Test Equipment	18
3	**The Base Station Subsystem**	**19**
3.1	Base Transceiver Station	19
3.1.1	Architecture and Functionality of a Base Transceiver Station	20
3.1.2	Base Transceiver Station Configurations	22
3.2	Base Station Controller	25
3.2.1	Architecture and Tasks of the Base Station Controller	26
3.3	Transcoding Rate and Adaptation Unit	28
3.3.1	Function of the Transcoding Rate and Adaptation Unit	28
3.3.2	Site Selection for Transcoding Rate and Adaptation Unit	28
3.3.3	Relationship Between the Transcoding Rate, Adaptation Unit, and Base Station Subsystem	29
4	**The Network Switching Subsystem**	**31**
4.1	Home Location Register and Authentication Center	32
4.2	Visitor Location Register	33
4.3	The Mobile-Services Switching Center	34
4.3.1	Gateway MSC	36
4.3.2	The Relationship Between MSC and VLR	36
4.4	Equipment Identity Register	37
5	**The OSI Reference Model**	**39**
5.1	Reasons for Standardization	39
5.2	Layering in the OSI Reference Model	40
5.3	Data Types of the OSI Reference Model	41
5.4	Information Processing in the OSI Reference Model	42

5.5	Advantages of the OSI Reference Model	42
5.6	The Seven Layers of the OSI Reference Model	43
5.6.1	Layer 1: The Physical Layer	43
5.6.2	Layer 2: The Data Link Layer	43
5.6.3	Layer 3: The Network Layer	44
5.6.4	Layer 4: The Transport Layer	44
5.6.5	Layer 5: The Session Layer	45
5.6.6	Layer 6: The Presentation Layer	45
5.6.7	Layer 7: The Application Layer	46
5.7	Comprehension Issues	46
5.7.1	An Analogy: The Move to Europe	47
6	**The Abis-Interface**	**51**
6.1	Channel Configurations	51
6.2	Alternatives for Connecting the BTS to the BSC	52
6.2.1	BTS Connection in a Serial Configuration	54
6.2.2	Connection of BTSs in Star Configuration	55
6.3	Signaling on the Abis-Interface	55
6.3.1	OSI Protocol Stack on the Abis-Interface	55
6.3.2	Layer 2	56
6.3.3	Layer 3	71
6.4	Bringing an Abis-Interface Into Service	87
6.4.1	Layer 1	87
6.4.2	Layer 2	87
7	**The Air-Interface of GSM**	**89**
7.1	The Structure of the Air-Interface in GSM	89
7.1.1	The FDMA/TDMA Scheme	89
7.1.2	Frame Hierarchy and Frame Numbers	90
7.1.3	Synchronization Between Uplink and Downlink	93
7.2	Physical Versus Logical Channels	94
7.3	Logical-Channel Configuration	94

7.3.1	Mapping of Logical Channels Onto Physical Channels	95
7.3.2	Possible Combinations	97
7.4	Interleaving	100
7.5	Signaling on the Air Interface	101
7.5.1	Layer 2 $LAPD_m$ Signaling	101
7.5.2	Layer 3	107
8	**Signaling System Number 7**	**125**
8.1	The SS7 Network	125
8.2	Message Transfer Part	126
8.3	Message Types in SS7	127
8.3.1	Fill-In Signal Unit	127
8.3.2	Link Status Signal Unit	128
8.3.3	Message Signal Unit	128
8.4	Addressing and Routing of Messages	130
8.4.1	Example: Determination of DPC, OPC, and SLS in a Hexadecimal Trace	131
8.4.2	Example: Commissioning of an SS7 Connection	132
8.5	Error Detection and Error Correction	133
8.5.1	Send Sequence Numbers and Receive Sequence Numbers (FSN, BSN, BIB, FIB)	135
8.5.2	BSN/BIB and FSN/FIB for Message Transfer	135
8.6	SS7 Network Management and Network Test	138
8.6.1	SS7 Network Test	139
8.6.2	Possible Error Cases	140
8.6.3	Format of SS7 Management Messages and Test Messages	142
8.6.4	Messages in SS7 Network Management and Network Test	142
9	**Signaling Connection Control Part**	**153**
9.1	Tasks of the SCCP	153

9.1.1	Services of the SCCP: Connection-Oriented Versus Connectionless	154
9.1.2	Connection-Oriented Versus Connectionless Service	154
9.2	The SCCP Message Format	156
9.3	The SCCP Messages	158
9.3.1	Tasks of the SCCP Messages	158
9.3.2	Parameters of SCCP Messages	159
9.3.3	Decoding a SCCP Message	167
9.4	The Principle of a SCCP Connection	167
10	**The A-Interface**	**171**
10.1	Dimensioning	171
10.2	Signaling Over the A-Interface	173
10.2.1	The Base Station Subsystem Application Part	173
10.2.2	The Message Structure of the BSSAP.	174
10.2.3	Message Types of the Base Station Subsystem Management Application Part	176
10.2.4	Decoding of a BSSMAP Message	183
11	**Transaction Capabilities and Mobile Application Part**	**185**
11.1	Transaction Capabilities Application Part	185
11.1.1	Addressing in TCAP	186
11.1.2	The Internal Structure of TCAP	187
11.1.3	Coding of Parameters and Data in TCAP	189
11.1.4	TCAP Messages Used in GSM	198
11.2	Mobile Application Part	208
11.2.1	Communication Between MAP and its Users	209
11.2.2	MAP Services	211
11.2.3	Local Operation Codes of the Mobile Application Part	214

11.2.4	Communication Between Application, MAP, and TCAP	220
12	**Scenarios**	**225**
12.1	Location Update	227
12.1.1	Location Update in the BSS	227
12.1.2	Location Update in the NSS	227
12.2	Equipment Check	227
12.3	Mobile Originating Call	233
12.3.1	Mobile Originating Call in the BSS	233
12.3.2	Mobile Originating Call in the NSS	233
12.4	Mobile Terminating Call	244
12.4.1	Mobile Terminating Call in the BSS	244
12.4.2	Mobile Terminating Call in the NSS	244
12.5	Handover	251
12.5.1	Measurement Results of BTS and MS	251
12.5.2	Analysis of a MEAS_RES/MEAS_REP	255
12.5.3	Handover Scenarios	256
13	**Quality of Service**	**275**
13.1	Tools for Protocol Measurements	275
13.1.1	OMC Versus Protocol Analyzers	276
13.1.2	Protocol Analyzer	278
13.2	Signaling Analysis in GSM	280
13.2.1	Automatic Analysis of Protocol Traces	280
13.2.2	Manual Analysis of Protocol Traces	284
13.3	Tips and Tricks	285
13.3.1	Identification of a Single Connection	285
13.4	Where in the Trace File to Find What Parameter?	287
13.5	Detailed Analysis of Errors on Abis Interface and A-Interface	287
13.5.1	Most Important Error Messages	291

| 13.5.2 | Error Analysis in the BSS | 296 |

Glossary — 303

About the Author — 405

Index — 407

1

Introduction

1.1 About This Book

Someone who wants to get to know the customs of a country frequently receives the advice to learn the language of that country. Why? Because the differences that distinguish the people of one country from those of another are reflected in the language. For example, the people of the islands of the Pacific do not have a term for *war* in their language. Similarly, some native tribes in the rain forests of the Amazon use up to 100 different terms for the color *green*.

The reflection of a culture in its language also applies to the area of computers. A closer look reveals that a modern telecommunications system, like the Global System for Mobile Communication (GSM), is nothing more than a network of computers. Depending on the application, a language has to be developed for such a communications network. That language is the signaling system, which allows intersystem communication by defining a fixed protocol. The study of the signaling system provides insight into the internal workings of a communication system.

The main purpose of this book, after briefly describing the GSM subsystems, is to lay the focus on the communications method—the signaling between these subsystems— and to answer questions such as which message is sent when, by whom, and why.

Because it is not always possible to answer all questions in a brief description or by analyzing signaling, details are covered in greater depth in the glossary. Furthermore, most of the items in the glossary contain references to GSM and International Telecommunication Union (ITU) Recommendations, which in turn allow for further research.

For the engineer who deals with GSM or its related systems on a daily basis, this book has advantages over other GSM texts in that it quickly gets to the point and can be used as a reference source. I hope the readers of this book find it helpful in filling in some of the gray areas on the GSM map.

1.2 Global System for Mobile Communication (GSM)

When the acronym GSM was used for the first time in 1982, it stood for *Groupe Spéciale Mobile*, a committee under the umbrella of *Conférence Européenne des Postes et Télécommunications* (CEPT), the European standardization organization.

The task of GSM was to define a new standard for mobile communications in the 900 MHz range. It was decided to use digital technology. In the course of time, CEPT evolved into a new organization, the European Telecommunications Standard Institute (ETSI). That, however, did not change the task of GSM. The goal of GSM was to replace the purely national, already overloaded, and thus expensive technologies of the member countries with an international standard.

In 1991, the first GSM systems were ready to be brought into so-called friendly-user operation. The meaning of the acronym GSM was changed that same year to stand for Global System for Mobile Communications. The year 1991 also saw the definition of the first derivative of GSM, the Digital Cellular System 1800 (DCS 1800), which more or less translates the GSM system into the 1800 MHz frequency range.

In the United States, DCS 1800 was adapted to the 1900 MHz band (Personal Communication System 1900, or PCS 1900). The next phase, GSM Phase 2, will provide even more end-user features than phase 1 of GSM did. In 1991, only "insiders" believed such a success would be possible because mobile communications could not be considered a mass market in most parts of Europe.

By 1992, many European countries had operational networks, and GSM started to attract interest worldwide. Time has brought substantial technological progress to the GSM hardware. GSM has proved to be a major commercial success for system manufacturers as well as for network operators.

How was such success possible? Particularly today, where Code Division Multiple Access (CDMA), Personal Handy Phone System (PHS), Digital Enhanced Cordless Telecommunications (DECT), and other systems try to mimic the success of GSM, that question comes to mind and is also discussed within the European standardization organizations.

The following factors were major contributors to the success of GSM:

- The liberalization of the monopoly of telecommunications in Europe during the 1990s and the resulting competition, which consequently lead to lower prices and more "market";
- The knowledge-base and professional approach within the Groupe Spéciale Mobile, together with the active cooperation of the industry;
- The lack of competition: For example, in the United States and Japan, competitive standards for mobile services started being defined only after GSM was already well established.

The future will show which system will prevail as the next generation of mobile communications. ETSI and the Special Mobile Group (SMG), renamed GSM, are currently standardizing the Universal Mobile Telecommunication System (UMTS). Japan is currently improving PHS.

The various satellite communications systems that now push into the market are another, possibly decisive, factor in providing mobile communications on a global basis.

1.2.1 The System Architecture of GSM: A Network of Cells

Like all modern mobile networks, GSM utilizes a cellular structure as illustrated in Figure 1.1.

The basic idea of a cellular network is to partition the available frequency range, to assign only parts of that frequency spectrum to any base transceiver station, and to reduce the range of a base station in order to reuse the scarce frequencies as often as possible. One of the major goals of network planning is to reduce interference between different base stations.

Anyone who starts thinking about possible alternatives should be reminded that current mobile networks operate in frequency ranges where attenuation is substantial. In particular, for mobile stations with low power emission, only small distances (less than 5 km) to a base station are feasible.

Besides the advantage of reusing frequencies, a cellular network also comes with the following disadvantages:

- An increasing number of base stations increases the cost of infrastructure and access lines.
- All cellular networks require that, as the mobile station moves, an active call is handed over from one cell to another, a process known as handover.

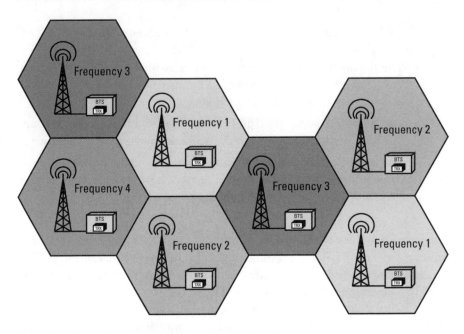

Figure 1.1 The radio coverage of an area by single cells.

- The network has to be kept informed of the approximate location of the mobile station, even without a call in progress, to be able to deliver an incoming call to that mobile station.
- The second and third items require extensive communication between the mobile station and the network, as well as between the various network elements. That communication is referred to as signaling and goes far beyond the extent of signaling that fixed networks use. The extension of communications requires a cellular network to be of modular or hierarchical structure. A single central computer could not process the amount of information involved.

1.2.2 An Overview on the GSM Subsystems

A GSM network comprises several elements: the mobile station (MS), the subscriber identity module (SIM), the base transceiver station (BTS), the base station controller (BSC), the transcoding rate and adaptation unit (TRAU), the mobile services switching center (MSC), the home location register (HLR), the visitor location register (VLR), and the equipment identity register (EIR). Together, they form a public land mobile network (PLMN). Figure 1.2 provides an overview of the GSM subsystems.

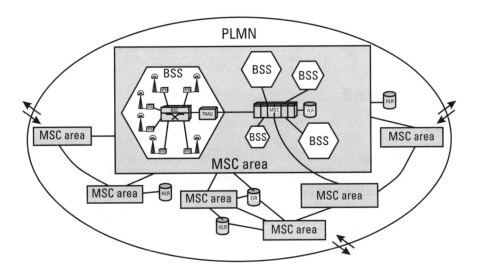

Figure 1.2 The architecture of a PLMN.

1.2.2.1 Mobile Station

GSM-PLMN contains as many MSs as possible, available in various styles and power classes. In particular, the handheld and portable stations need to be distinguished.

1.2.2.2 Subscriber Identity Module

GSM distinguishes between the identity of the subscriber and that of the mobile equipment. The SIM determines the directory number and the calls billed to a subscriber. The SIM is a database on the user side. Physically, it consists of a chip, which the user must insert into the GSM telephone before it can be used. To make its handling easier, the SIM has the format of a credit card or is inserted as a plug-in SIM. The SIM communicates directly with the VLR and indirectly with the HLR.

1.2.2.3 Base Transceiver Station

A large number of BTSs take care of the radio-related tasks and provide the connectivity between the network and the mobile station via the Air-interface.

1.2.2.4 Base Station Controller

The BTSs of an area (e.g., the size of a medium-size town) are connected to the BSC via an interface called the Abis-interface. The

BSC takes care of all the central functions and the control of the subsystem, referred to as the base station subsystem (BSS). The BSS comprises the BSC itself and the connected BTSs.

1.2.2.5 Transcoding Rate and Adaptation Unit

One of the most important aspects of a mobile network is the effectiveness with which it uses the available frequency resources. Effectiveness addresses how many calls can be made by using a certain bandwidth, which in turn translates into the necessity to compress data, at least over the Air-interface. In a GSM system, data compression is performed in both the MS and the TRAU. From the architecture perspective, the TRAU is part of the BSS. An appropriate graphical representation of the TRAU is a black box or, more symbolically, a clamp.

1.2.2.6 Mobile Services Switching Center

A large number of BSCs are connected to the MSC via the A-interface. The MSC is very similar to a regular digital telephone exchange and is accessed by external networks exactly the same way. The major tasks of an MSC are the routing of incoming and outgoing calls and the assignment of user channels on the A-interface.

1.2.2.7 Home Location Register

The MSC is only one subcenter of a GSM network. Another subcenter is the HLR, a repository that stores the data of a large number of subscribers. An HLR can be regarded as a large database that administers the data of literally hundreds of thousands of subscribers. Every PLMN requires at least one HLR.

1.2.2.8 Visitor Location Register

The VLR was devised so that the HLR would not be overloaded with inquiries on data about its subscribers. Like the HLR, a VLR contains subscriber data, but only part of the data in the HLR and only while the particular subscriber roams in the area for which the VLR is responsible. When the subscriber moves out of the VLR area, the HLR requests removal of the data related to a subscriber from the VLR. The geographic area of the VLR consists of the total area covered by those BTSs that are related to the MSCs for which the VLR provides its services.

1.2.2.9 Equipment Identity Register

The theft of GSM mobile telephones seems attractive, since the identities of subscribers and their mobile equipment are separate. Stolen equipment can be reused simply by using any valid SIM. Barring of a subscriber by the operator does not bar the mobile equipment. To prevent that kind of misuse, every GSM terminal equipment contains a unique identifier, the international mobile equipment identity (IMEI). It lies within the realm of responsibilities of a network operator to equip the PLNM with an additional database, the EIR, in which stolen equipment is registered and so can be used to bar fraudulent calls and even, theoretically, to track down a thief (by analyzing the related SIM data).

1.3 The Focus of This Book

This book describes briefly the GSM subsystems, their structure, and their tasks. However, the focus of this book lies not on the GSM network elements themselves but on the interfaces between them.

Among others, the following issues will be addressed:

- What signaling standards and what protocols are used to serve connection requests by mobile subscribers?
- How are the various interfaces utilized?
- What happens in case of errors?
- Although GSM uses available signaling standards, where are the GSM specific adaptations?

One has to remember that most of the signaling is necessary to support the mobility of a subscriber. All messages of the area mobility management (MM) and radio resource management (RR), in particular, serve only that purpose. Only a fraction of the exchanged messages are used for the connection setup as such, and those are all the messages that are related to call control (CC).

A presentation of the Open System Interconnection (OSI) Reference Model is mandatory in a book in which the focus is on signaling.

Another focus of the text is on the application of the various protocols for error analysis. Which error indication is sent by the system and when? How is such an indication interpreted? What are the potential sources of errors?

A word on coding of parameters and messages should be added here: Coding of message types and other essential parameters are always included. However, because this book has no intention of being a copy of the GSM Recommendations, it deliberately refrains from providing a complete list of all parameters of all interfaces.

The value of protocol test equipment for error analysis and routine testing is indisputably high, but what help do programs for automatic analysis provide? Those questions will be answered as well.

A large part of this book is taken up by a glossary, which provides descriptions of all abbreviations, terms, and processes that a reader may confront during work on GSM.

1.4 Signaling

The main focus of this book is on the signaling between the various network elements of GSM. The questions arise of what signaling actually constitutes and what it is used for. Although we do not want to go back to the basics of telecommunications to answer those questions, a number of basic explanations do seem necessary.

1.4.1 What is Signaling?

Signaling is the language of telecommunications that machines and computers use to communicate with each other. In particular, the signals that a user enters need to be converted to a format that is appropriate for machines and then transmitted to a remote entity. The signals (e.g., the identity of a called party) are not part of the communication as such, that is, they are not a payload or a revenue-earning entity.

Signaling is comparable to the pilots and the flight attendants on an airplane. The crew members are no "payload," but they are necessary to carry the payload. Another, perhaps more appropriate illustration is to consider the now almost extinct telephone operator, whose function it was to carry out the signaling function and switching of a telecommunications system by connecting cables between the appropriate incoming and outgoing lines.

1.4.2 How is Signaling Performed?

When calls were set up manually, signaling consisted mostly of direct current impulses, which allowed a central office to determine the dialed digits. Some readers may still remember the rotary telephone sets, in which the impulses were created mechanically by the spin of the rotor. The situation changed completely with the entry of computer technology into telecommunications. The microchip utilized by telecommunications opens today, at the end of the twentieth century, a multitude of new signaling functionality, which were unthinkable even 20 years ago. Computers are the backbone of modern telecommunications systems.

This new technology makes mobile communications possible in the first place. The signaling requirements of modern mobile systems are so vast that the former technology would not have been able to manage them. Computerization, however, did not change much of the principle. As in the old days, electrical or optical signals are sent, over an appropriate medium (typically serially) and interpreted by the receiver. What did change is the speed of the transmission. The progress in this area has been exponential.

The smallest unit of a signal is called a *bit* and can, for example, be represented by an electric voltage, which a receiver can measure during a specified period of time. If the receiver measures the voltage as "low" over the specified time period, it interprets the value as a 0. If the voltage is "high," the receiver interprets the value as a 1. It does not matter which level represents which value, so long as both the sender and the receiver agree on which is which.

A sequence of bits allows the coding and sending of complex messages, which, in turn, allows a process to be controlled or information to be conveyed. The result is a bit stream, as shown in Figure 1.3.

Pulse code modulation (PCM) is the worldwide process for transmission of digital signals. PCM is used to transmit both signaling data and payload. PCM is categorized into hierarchies, depending on the transmission rate. The PCM link of 2 megabits per second (Mbps) (one that is referred to frequently in this book) is only one variant of many. By utilizing a time-division

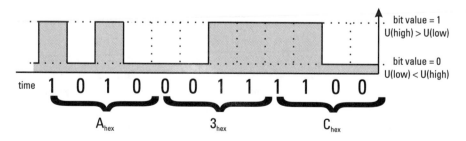

Figure 1.3 Decoding of a bit stream.

multiplexing technique, such a 2-Mbps PCM link can, among others, be partitioned into 32 independent channels, each capable of carrying 64 kilobits per second (Kbps).

Another aspect of the change that the digital technology has enabled reveals its advantage only after a second look. Almost all signaling standards, like Signaling System Number 7 (SS7) and Link Access Protocol for the D-channel (LAPD) separate the traffic channel from the signaling or control channel. This is referred to as outband signaling, in contrast to inband signaling. In the case of inband signaling, all the control information is carried within the traffic channel. Although outband signaling requires the reservation of a traffic channel, it makes a more efficient use of resources overall. The reason for that lies in the reduced occupation time of the traffic channel, which is not needed during call setup. Both call setup and call release can be carried out for many connections via one control channel, since signaling data use the resources more economically. One 64-Kbps time slot out of a 2-Mbps PCM link typically is used for signaling data; a call setup consumes about 1 to 2 Kbps.

1.4.3 What is Signaling Used For?

The main task of signaling is still to set up and to clear a connection between end users or machines. Today, constantly new applications are added. Among them are automated database accesses, in which telecommunications systems call each other and which are fairly transparent to a caller, or the wide area of supplementary services, of which only call forwarding is mentioned here as an example. The glossary provides a list of all GSM supplementary services.

1.5 Representation of Messages

When working with protocol test equipment and in practical work, message names usually are abbreviated. Most GSM and ITU Telecommunication Standardization Sector (ITU-T) Recommendations list the well-defined abbreviations and acronyms, which this book also uses to a large extent. The complete message names and explanations can be found in the respective chapters.

Since a picture often expresses more than a thousand words, this book contains a large number of figures and protocol listings. The various messages illustrated in the figures show parameters, which are formatted per interface and are presented as shown in Figures 1.4(a) through 1.4(e).

Introduction 11

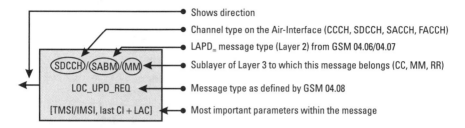

Figure 1.4(a) Format for messages over the Air-interface (LAPD$_m$, GSM 04.08).

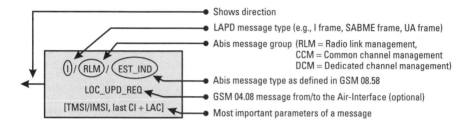

Figure 1.4(b) Format for messages over the Abis-interface (LAPD, GSM 08.58).

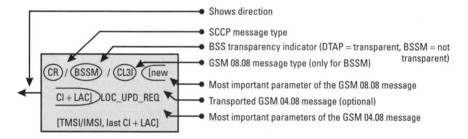

Figure 1.4(c) Format for messages over the A-interface [SS7, signaling connection control part (SCCP), GSM 08.06, GSM 08.08].

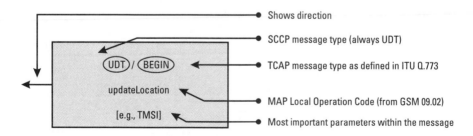

Figure 1.4(d) Format for mobile application part (MAP) messages over all network switching subsystem (NSS) interfaces [SS7, SCCP, transaction capabilities application part (TCAP), MAP].

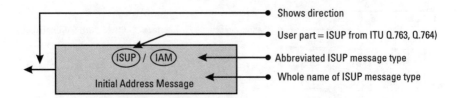

Figure 1.4(e) Format for ISUP messages between MSCs and toward the Integrated Services Digital Network (ISDN) [SS7 and the ISDN user part (ISUP)].

2

The Mobile Station and the Subscriber Identity Module

The GSM telephone set and the SIM are the only system elements with which most users of GSM have direct contact. The GSM telephone set and the SIM form an almost complete GSM system within themselves with all the functionality, from ciphering to the HLR. Figure 2.1 shows a block diagram of a mobile station with a SIM slot.

2.1 Subscriber Identity Module

The SIM is a microchip that is planted on either a check card (ID-1 SIM) or a plastic piece about 1 cm square (plug-in SIM). Figure 2.2 shows both variants. Except for emergency calls, a GSM mobile phone cannot be used without the SIM. The GSM terminology distinguishes between a mobile station and mobile equipment. The mobile equipment becomes a mobile station when the SIM is inserted. There is no difference in functionality between the ID-1 SIM and the plug-in SIM, except for size, which is an advantage for the plug-in SIM when used in a small handheld telephone. Today, many network operators offer (at an additional cost) identical pairs of ID-1 SIM/plug-in SIM, so the same SIM can be used in a car phone and in a handheld telephone.

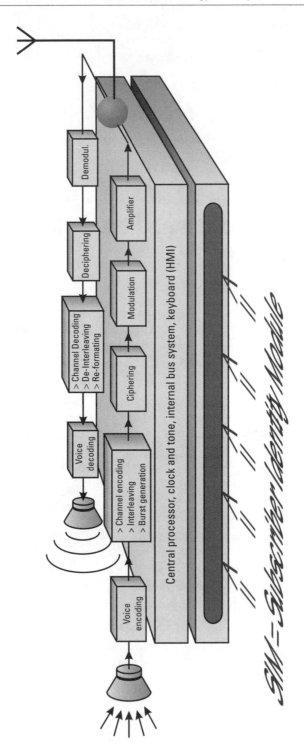

Figure 2.1 Block diagram of a GSM MS.

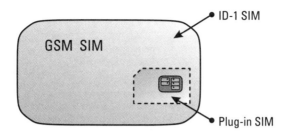

Figure 2.2 Plug-in SIM and ID-1 SIM.

2.1.1 The SIM as a Database

The major task of a SIM is to store data. That does not mean that the data is only subscriber data. One has to differentiate between data types for various tasks. The most important parameters that a SIM holds are presented in Table 2.1. It should be noted that the list is not complete and that the SIM can also be used to store, for example, telephone numbers.

2.1.2 Advantage for the Subscriber

The SIM is one of the most interesting features for a user of GSM, because it permits separation of GSM telephone equipment and the related database. In other words, the subscriber to a GSM system is not determined by the identity of the mobile equipment but by the SIM, which always has to be inserted into the equipment before it can be used. This is the basis for personal mobility.

Table 2.1
Data Stored on a SIM

Parameter	Remarks
Administrative data	
PIN/PIN2 (m/v)	Personal identification number; requested at every powerup (PIN or PIN2)
PUK/PUK2 (m/f)	PIN unblocking key; required to unlock a SIM
SIM service table (m/f)	List of the optional functionality of the SIM
Last dialed number) (o/v)	Redial
Charging meter (o/v)	Charges and time increments can be set
Language (m/v)	Determines the language for prompts by the mobile station

Table 2.1 (continued)

Parameter	Remarks
Security related data	
Algorithm A3 and A8 (m/f)	Required for authentication and to determine Kc
Key Ki (m/f)	Individual value; known only on SIM and the HLR
Key Kc (m/v)	Result of A8, Ki, and random number (RAND)
CKSN (m/v)	Ciphering key sequence number
Subscriber data	
IMSI (m/f)	International mobile subscriber identity
MSISDN (o/f)	Mobile subscriber ISDN; directory number of a subscriber
Access control class(es) (m/f)	For control of network access
Roaming data	
TMSI (m/v)	Temporary mobile subscriber identity
Value of T3212 (m/v)	For location updating
Location updating status	Is a location update required?
LAI (m/v)	Location area information
Network color codes (NCCs) of restricted PLMNs (m/v)	Maximum of 4 PLMNs can be entered on a SIM after unsuccessful location update; cause "PLMN not allowed." Oldest entry deleted when more than 4 restricted PLMNs are found.
NCCs of preferred PLMNs (o/v)	What PLMN should the MS select, if there is more than one to choose from and the home PLMN is not available?
PLMN data	
NCC, mobile country code (MCC), and mobile network code (MNC) of the home PLMN (m/f)	Network identifier
Absolute radio frequency channel numbers (ARFCNs) of home PLMN (m/f)	Frequencies for which the home PLMN is licensed.

Legend: m = mandatory; o = optional; f = fixed, unchangeable value; v = changeable

Because of the SIM, a GSM customer can use different telephones (e.g., a car phone and a handheld phone) and still be reachable under the same directory number. Even in case of a defect in the user's GSM telephone, any other GSM telephone can be used instead, simply by changing the SIM.

2.2 Mobile Station

A GSM terminal is, even for experts, a technical marvel. Consider the rate at which prices have fallen, the complexity of the devices, and the large number of different types of equipment available. All the functionality known from the BTS transmitter/receiver (TRX), like Gaussian minimum shift keying (GMSK) modulation/demodulation up to channel coding/decoding, also needs to be implemented in an MS.

Other MS-specific functionalities need to be mentioned, like dual-tone multifrequency (DTMF) generation and the most important issue, the economical use of battery power.

From the perspective of the protocol, the MS is not only a peer of the BTS but communicates directly with the MSC and the VLR, via the mobility management (MM) and call control (CC). Furthermore, the MS has to be able to provide a transparent interface (terminal adaptation function, or TAF) for data and fax connections to external devices.

2.2.1 Types of Mobile Stations

The most common way to distinguish among GSM mobile equipment is by the power class ratings, in which the value specifies the maximum transmission power of an MS.

When GSM was introduced, five power classes were defined for GSM 900, of which the most powerful class allowed for a 20W output. That class is no longer supported; currently, the most powerful rating is 8W. The power emission of DCS 1800 and PCS 1900 mobiles is generally lower. The Glossary lists the power classes for all three standards.

2.2.2 Functionality

GSM Recommendation 02.07 describes in detail what functionality mobile equipment has to support and what features are optional. The most important and mandatory features are:

- DTMF capability;
- Short-message service (SMS) capability;
- Availability of the ciphering algorithms A5/1 and A5/2;
- Display capability for short messages, dialed numbers, and available PLMN;
- Support of emergency calls, even without the SIM inserted;
- "Burned-in" IMEI.

2.2.3 Mobile Stations as Test Equipment

An MS is a useful test tool for the laboratory testing of a new network function. Several manufacturers offer, for that purpose, a semistationary MS, which allows manipulation of specific system parameters, to test the behavior of new software or hardware.

Besides those complex and expensive pieces of equipment used mainly in laboratories, a number of standard mobile telephones exist, which can easily be modified with additional packages to act as mobile test equipment. Such equipment is connected to a personal computer and uses standard functionality to monitor signaling between the network and the MS. Usually, it is also able to represent the test results in tabular or graphical form.

Despite those advantages, the test mobile stations seldom are used for protocol and error analysis, because the results are not representative from a statistical point of view and can be gathered only with substantial effort and time.

Nonetheless, special test mobiles are necessary tools for network operators, to monitor coverage and evaluate the behavior of handover as a customer would experience it.

3

The Base Station Subsystem

Via the Air-interface, the BSS provides a connection between the MSs of a limited area and the network switching subsystem (NSS). The BSS consists of the following elements:

- One or more BTSs (base tranceiver station);
- One BSC (base station controller);
- One TRAU (transcoding rate and adaptation unit).

The tasks and the structure of those elements or modules are described in this chapter.

3.1 Base Transceiver Station

The BTS provides the physical connection of an MS to the network in form of the Air-interface. On the other side, toward the NSS, the BTS is connected to the BSC via the Abis-interface.

The manufacturers of BTS equipment have been able to reduce its size substantially. The typical size in 1991 was that of an armoire; today the size is comparable to that of a mailbox. The basic structure of the BTS, however, has not changed. The block diagram and the signal flow of a BTS with one TRX are shown in Figure 3.1. The GSM Recommendations allow for one BTS to host up to 16 TRXs. In the field, the majority of the BTSs host between one and four TRXs.

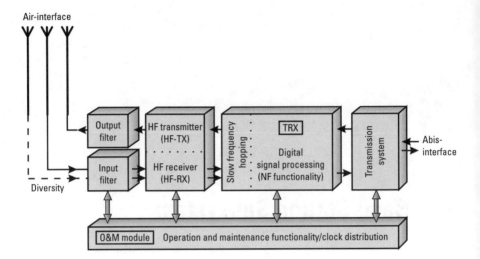

Figure 3.1 Block diagram of a BTS with one TRX.

3.1.1 Architecture and Functionality of a Base Transceiver Station

3.1.1.1 Transmitter/Receiver Module

The TRX module is, from the perspective of signal processing, the most important part of a BTS. The TRX consists of a low-frequency part for digital signal processing and a high-frequency part for GMSK modulation and demodulation. Both parts are connected via a separate or an integrated frequency hopping unit. All other parts of the BTS are more or less associated with the TRXs and perform auxiliary or administrative tasks.

A TRX with integrated frequency hopping serves the tasks listed in Table 3.1.

3.1.1.2 Operations and Maintenance Module

The operations and maintenance (O&M) module consists of at least one central unit, which administers all other parts of the BTS. For those purposes, it is connected directly to the BSC by means of a specifically assigned O&M channel. That allows the O&M module to process the commands from the BSC or the MSC directly into the BTS and to report the results. Typically, the central unit also contains the system and operations software of the TRXs. That allows it to be reloaded when necessary, without the need to "consult" the BSC. Furthermore, the O&M module provides a human-machine interface (HMI), which allows for local control of the BTS.

Table 3.1
Tasks of a TRX With Integrated Frequency Hopping

Function	LF	HF
Channel coding and decoding	●	
Interleaving and ordering again	●	
Encryption and decryption (ciphering)	●	
Slow frequency hopping	●	
Burst formatting	●	
TRAU frame formatting and conversion in direction to/from the BSC, setup of the LAPD connection to the BSC	●	
GMSK modulation of all downlink data	●	●
GMSK demodulation of all received MS signals	●	●
Creation and transmission of the broadcast common control channel (BCCH) on channel 0 of the BCCH-TRX	●	●
Measurement of signal strength and quality for active connections Provision of the results to the BSC (MEAS_RES message)	●	●
Interference measurements (idle channel measurements) on free channels and forwarding of the results to the BSC in a RF_RES_IND message	●	●

LF = low frequency part of the TRX; HF = high frequency part of the TRX.

3.1.1.3 Clock Module

The modules for clock generation and distribution also are part of the O&M area. Although the trend is to derive the reference clock from the PCM signal on the Abis-interface, a BTS internal clock generation is mandatory. It is especially needed when a BTS has to be tested in a standalone environment, that is, without a connection to a BSC or when the PCM clock is not available due to link failure.

Still, there is a cost savings benefit in the approach of deriving the clock from the PCM signal. By doing so, much cheaper internal clock generators can be applied, because they do not require the same long-term stability as an independent clock generator. Besides, there is no need for frequent maintenance checks on the clock modules, since they synchronize themselves with the clock coming from the PCM link.

When analyzing errors in call handling, particularly in the area of handover, even minor deviations from the clock have to be considered as possible causes for errors. GSM requires that all the TRXs of a BTS use the same clock signal. The accuracy of the signal has to have a precision of at least 0.05 parts per million (ppm). For example, a clock generator that derives the clock from a 10 MHz signal has to be able to provide a clock with a frequency accuracy of 10 MHz ±0.5 Hz ($10 \cdot 10^6$ Hz $\cdot 0.05 \cdot 10^{-6}$ = 0.5 Hz).

3.1.1.4 Input and Output Filters

Both input and output filters are used to limit the bandwidth of the received and the transmitted signals. The input filter typically is a nonadjustable wideband filter that lets pass all GSM 900, all DCS 1800, or all PCS 1900 frequencies in the uplink direction. In contrast, remote-controllable filters or wideband filters are used for the downlink direction that limits the bandwidth of the output signal to 200 kHz. When necessary, the O&M center (OMC) controls the settings of the filters, as in the case of a change in frequency.

3.1.2 Base Transceiver Station Configurations

Different BTS configurations, depending on load, subscriber behavior, and morph structure, have to be considered to provide optimum radio coverage of an area. The most important BTS configurations of a BTS are presented next.

3.1.2.1 Standard Configuration

All BTSs are assigned different cell identities (CIs). A number of BTSs (in some cases, a single BTS) form a location area. Figure 3.2 shows three location areas with one, three, and five BTSs. The systems are usually not fine-synchronized (see *synchronized handover* in the Glossary), which prevents synchronized handover between them. That method of implementing BTSs is the one most frequently used. For urban areas with growing traffic density, that may change soon. For this situation, the configurations described in Sections 3.1.2.2 and 3.1.2.3 are more appropriate.

3.1.2.2 Umbrella Cell Configuration

The umbrella cell configuration consists of one BTS with high transmission power and an antenna installed high above the ground that serves as an "umbrella" for a number of BTSs with low transmission power and small diameters (Figure 3.3).

Such a configuration appears to make no sense at first, because the frequency of the umbrella cell can not be reused in all the cells of that area due to interference. Interference even over a large distance was one of the reasons why the high radio and television towers were abandoned as sites for antennas shortly after they were brought into service at the initial network startup.

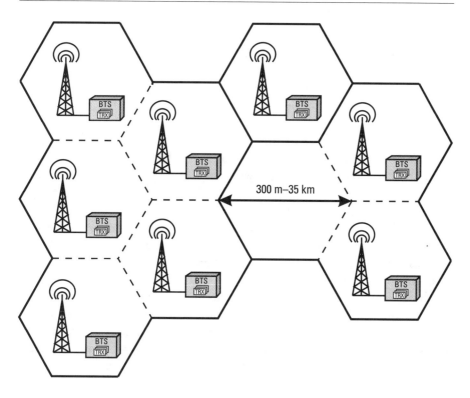

Figure 3.2 BTSs in standard configuration.

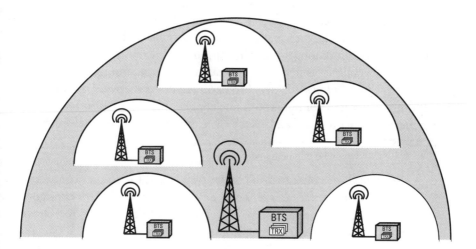

Figure 3.3 Umbrella cell with five smaller cells.

The umbrella cell configuration still has its merits in certain situations and therefore may result in relief from load and an improvement of the network. For example, when cars are moving at rather high speeds through a network of small cells, almost consecutive handovers from one cell to the next are necessary to maintain an active call. This situation is applicable in every urban environment that features city highways. Consequently, the handovers result in a substantial increase of the signaling load for the network as well as in an unbearable signal quality degradation for the end user. On the other hand, small cells are required to cope with the coverage demand in an urban environment.

The way out of this dilemma is to use both large and small cells at the same time, that is, the umbrella cell configuration. The umbrella cell can be protected from overload when traffic from only fast-moving users is assigned to it. This, on the other hand, reduces the signaling load of the small cells and improves the signal quality for the fast-moving traffic. The speed of a user can be determined to sufficient accuracy by the change of the timing advance (TA) parameter. Its value is updated in the BSC every 480 milliseconds (ms) by means of the data provided in the MEAS_RES message. The BSC decides whether to use the umbrella cell or one of the small cells. GSM has not specified the umbrella cell configuration, which requires additional functionality in the BSC, a manufacturer's proprietary function.

3.1.2.3 Sectorized (Collocated) Base Transceiver Stations

The term sectorized, or collocated, BTSs refers to a configuration in which several BTSs are collocated at one site but their antennas cover only an area of 120 or 180 degrees. Figure 3.4 illustrates the concept. Typically, it is implemented with BTSs with few TRXs and low transmission power. Like the umbrella cell configuration, this configuration is used mostly in highly populated areas. A peculiarity is that it is fairly easy to fine-synchronize the cells with each other, which allows for synchronized handover between them. Even though in a collocated configuration, one channel per BTS has to be used for the generation of the BCCH, such a configuration has the following advantages:

- Sectorized, or collocated, BTSs are well suited for a serial connection of the Abis-interface (discussed in detail in Chapter 6). This configuration has the potential to save costs for access lines to the BSC. Otherwise, multiple sites require multiple (leased) lines.

- From the radio perspective, the advantage of using cells with a 120-degree angle is that it allows reuse of frequencies in one sector

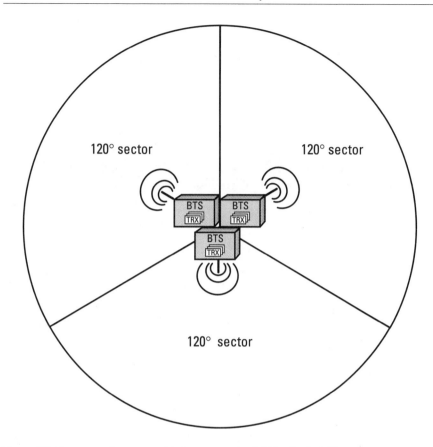

Figure 3.4 Coverage of an area with three sectorized BTSs. Each BTS covers a segment of 120 degrees.

(one direction), which otherwise would cause interference with neighbor cells if an omnidirectional cell were used.

- Sectorization eases the demand for frequencies, particularly in an urban environment.

3.2 Base Station Controller

The BSC forms the center of the BSS. A BSC can, depending on the manufacturer, connect to many BTSs over the Abis-interface. The BSC is, from a technical perspective, a small digital exchange with some mobile-specific extensions. The BSC was defined with the intention of removing most of the radio-related load from the MSC. The BSC's architecture and its tasks are a consequence of that goal. For simplicity, Figure 3.5 uses the same hardware for both the Abis-interface and the A-interface, which is not a requirement.

3.2.1 Architecture and Tasks of the Base Station Controller

3.2.1.1 Switch Matrix

Because the BSC has the functionality of a small digital exchange, its function is to switch the incoming traffic channels (A-interface from the MSC) to the correct Abis-interface channels. The BSC, therefore, comes with a switch matrix that (1) takes care of the relay functionality and (2) can be used as the internal control bus.

3.2.1.2 Terminal Control Elements of the Abis-Interface

The connection to the BTSs is established via the Abis-terminal control elements (TCEs), which, more or less independently from the BSC's central unit, provide the control function for a TRX or a BTS. The number of Abis TCEs that a BSC may contain depends largely on the number of BTSs and on the system manufacturer.

Major tasks of the Abis-TCEs are to set up LAPD connections toward the BTS peers, the transfer of signaling data, and last—but not least—the transparent transfer of payload.

Depending on the manufacturer, the Abis TCEs also may be responsible for the administration of BTS radio resources. That entails the assignment and

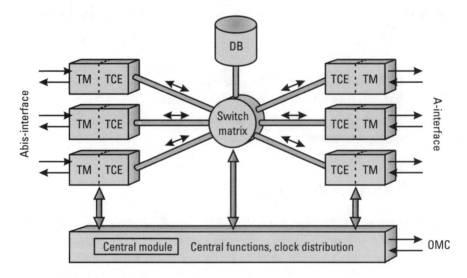

Figure 3.5 Block diagram of a BSC.

release of signaling and traffic channels over the Abis-interface and the Air-interface and for the evaluation of measurement results from the BTS concerning busy and idle channels, which are relevant for power control and used in making decisions about handovers. The final control functionality always remains with the BSC, although GSM explicitly allows the BTS to preprocess the measurement results. Depending on the manufacturer, those functions also can be assumed or controlled by a central unit.

Connections from the Abis TCEs to the A-TCEs are realized by the switch matrix. On the other side, the PCM connections are achieved by associated transmission elements.

3.2.1.3 A-Interface Terminal Control Elements

The connection of a BSC to the MSC is established via the A-TCEs. Although every BSC is connected to only one MSC, a large number of A-TCEs is needed to support the A-interface, since all the payload and the major part of the signaling data of the entire BSS have to be conveyed over this interface.

Among the tasks of some, but usually not all A-TCEs is setting up and operating the SS7/SCCP connection toward the MSC. The number of necessary signaling channels depends largely on the predicted traffic load (see also Chapter 10).

3.2.1.4 Database

The BSC is the control center of the BSS. In that capacity, the BSC must maintain a relatively large database in which the maintenance status of the whole BSS, the quality of the radio resources and terrestrial resources, and so on are dynamically administrated. Furthermore, the BSC database contains the complete BTS operations software for all attached BTSs and all BSS specific information, such as assigned frequencies.

3.2.1.5 Central Module

One of the major tasks of the BSC is to decide when a handover should take place. The BSC may decide on intra-BTS handover and intra-BSC handover without needing the MSC. In contrast, for all BSC external handovers, the BSC needs to involve the MSC. Handover decision and power control are main tasks of the central module.

3.2.1.6 Connection to the OMC

Another functionality that many manufacturers have decided the central module should perform is the connection to the OMC. Every BSS is supervised and managed by an OMC via the BSC.

3.3 Transcoding Rate and Adaptation Unit

3.3.1 Function of the Transcoding Rate and Adaptation Unit

One of the most interesting functions in GSM involves the TRAU, which typically is located between the BSC and the MSC. The task of the TRAU is to compress or decompress speech between the MS and the TRAU. The used method is called regular pulse excitation–long term prediction (RPE-LTP). It is able to compress speech from 64 Kbps to 16 Kbps, in the case of a fullrate channel (net bit rate with fullrate is 13 Kbps) and to 8 Kbps in the case of a halfrate channel (net bit rate with halfrate is 6.5 Kbps).

Note that the TRAU is not used for data connections.

3.3.2 Site Selection for Transcoding Rate and Adaptation Unit

Although speech compression is intended mainly to save resources over the Air-interface, it also is suitable to save line costs when applied on terrestrial links, as illustrated schematically in Figure 3.6. When the TRAU is installed at the MSC site (see top portion of Figure 3.6), a fullrate speech channel uses only 16 Kbps over the link from the BSC to the MSC.

The specifications allow for the installation of the TRAU between the BTS and the BSC. That requires, however, the use of 64-Kbps channels between the BSC and the MSC and hence the use of more links (see bottom portion of Figure 3.6).

This variant is, therefore, used only infrequently. In fact, most of the time, the TRAU is installed at the site of the MSC to get the most benefit from the compression.

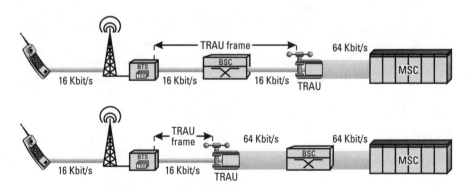

Figure 3.6 Possible sites for the TRAU in the signal chain.

3.3.3 Relationship Between the Transcoding Rate and Adaptation Unit, and Base Station Subsystem

The TRAU is functionally assigned to the BSS, independently of where it actually is located. The reason for that is the following.

Both the BTS and the TRAU have an interface for payload that is transparent for the BSC. The payload is formatted in TRAU frames, then transparently sent over PCM links between the TRAU and the BTS in cycles of 20 ms. That applies to both directions. The data contained in the TRAU frames form the input and output values for channel coding.

For data connections, the compression functionality has to be switched off. The type of connection (data/speech) is communicated to the TRAU during the assignment of the traffic channel. As illustrated in Figure 3.7, the BTS starts to transmit TRAU frames in the uplink, immediately after receiving the CHAN_ACT message. Those TRAU frames carry inband signaling, which is exchanged between the BTS-TRX (or more precisely the coding unit) and the TRAU, to consolidate the characteristics of a connection. Part of the control information is, in particular, synchronization data, discontinuous transmission (DTX) on/off, and the connection type (halfrate/fullrate).

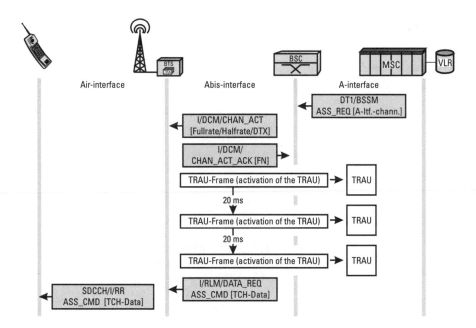

Figure 3.7 Activation of the TRAU during assignment of a traffic channel.

Note that TRAU frames are sent over traffic channels and not over the associated control channels and hence are transparent to protocol analysis. The TRAU frames are, nevertheless, very important for error analysis on data connections.

4

The Network Switching Subsystem

The NSS plays the central part in every mobile network. While the BSS provides the radio access for the MS, the various network elements within the NSS assume responsibility for the complete set of control and database functions required to set up call connections using one or more of these features: encryption, authentication, and roaming. To satisfy those tasks, the NSS consists of the following:

- MSC (mobile switching center);
- HLR (home location register)/authentication center (AuC);
- VLR (visitor location register);
- EIR (equipment identity register).

The subsystems are interconnected directly or indirectly via the worldwide SS7 network. The network topology of the NSS is more flexible than the hierarchical structure of the BSS. Several MSCs may, for example, use one common VLR; the use of an EIR is optional, and the required number of subscribers determines the required number of HLRs.

Figure 4.1 provides an overview of the interfaces between the different network elements in the NSS. Note that most interfaces are virtual, that is, they are defined as reference points for signaling between the network elements.

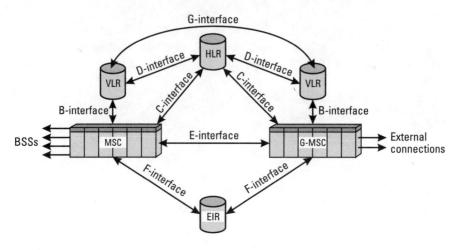

Figure 4.1 The NSS.

4.1 Home Location Register and Authentication Center

Every PLMN requires access to at least one HLR as a permanent store of data. The concept is illustrated in Figure 4.2. The HLR can best be regarded as a large database with access times that must be kept as short as possible. The faster the response from the database, the faster the call can be connected. Such a database is capable of managing data for literally hundreds of thousands subscribers.

Within the HLR, subscriber-specific parameters are maintained, such as the parameter K_i, which is part of security handling. It is never transmitted on any interface and is known only to the HLR and the SIM, as shown in Figure 4.2.

Each subscriber is assigned to one specific HLR, which acts as a fixed reference point and where information on the current location of the user is stored. To reduce the load on the HLR, the VLR was introduced to support the HLR by handling many of the subscriber-related queries (e.g., localization and approval of features).

Because of the central function of the HLR and the sensitivity of the stored data, it is essential that every effort is taken to prevent outages of the HLR or the loss of subscriber data.

The AuC is always implemented as an integral part of the HLR. The reason for this is that although GSM mentions the interface between the AuC and the HLR and has even assigned it a name, the H-interface, it was never specified in sufficient detail to be a standalone entity. The only major function assigned to the AuC is to calculate and provide the authentication-triplets, that

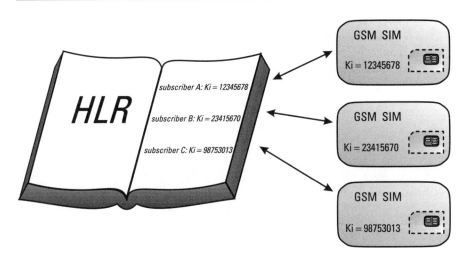

Figure 4.2 Only the SIM and the HLR know the value of K_i.

is, the signed response (SRES), the random number (RAND), and K_c. For each subscriber, up to five such triplets can be calculated at a time and sent to the HLR. The HLR, in turn, forwards the triplets to the VLR, which uses them as input parameters for authentication and ciphering.

The Glossary provides a detailed description of the authentication procedure.

4.2 Visitor Location Register

The VLR, like the HLR, is a database, but its function differs from that of the HLR While the HLR is responsible for more static functions, the VLR provides dynamic subscriber data management. Consider the example of a roaming subscriber. As the subscriber moves from one location to another, data are passed between the VLR of the location the subscriber is leaving ("old" VLR) to the VLR of the location being entered ("new" VLR). In this scenario, the old VLR hands over the related data to the new VLR. There are times when the new VLR has to request the subscriber's HLR for additional data.

This question then arises: Does the HLR in GSM assume responsibility for the management of those subscribers currently in its geographic area? The answer is no. Even if the subscriber happens to be in the home area, the VLR of that area handles the dynamic data. This illustrates another difference between the HLR and the VLR. The VLR is assigned a limited geographical area, while the HLR deals with tasks that are independent of a subscriber's location. The term *HLR area* has no significance in GSM, unless it refers to the whole PLMN. Typically, but not necessarily, a VLR is linked with a single MSC. The GSM standard allows, as Figure 4.3 illustrates, the association of one VLR with several MSCs.

The initial intentions were to specify the MSC and the VLR as independent network elements. However, when the first GSM systems were put into service in 1991, numerous deficiencies in the protocol between the MSC and the VLR forced the manufacturers to implement proprietary solutions. That is the reason the interface between the MSC and the VLR, the B-interface, is not mentioned in the specifications of GSM Phase 2. GSM Recommendation 09.02 now provides only some basic guidelines on how to use that interface.

Table 4.1 lists the most important data contained in the HLR and the VLR.

4.3 The Mobile-Services Switching Center

From a technical perspective, the MSC is just an ordinary Integrated Services Digital Network (ISDN) exchange with some modifications specifically required to handle the mobile application. That allows suppliers of GSM systems to offer their switches, familiar in many public telephone networks, as MSCs. SIEMENS with its EWSD technology and ALCATEL with the S12 and the E10 are well-known examples that benefit from such synergy.

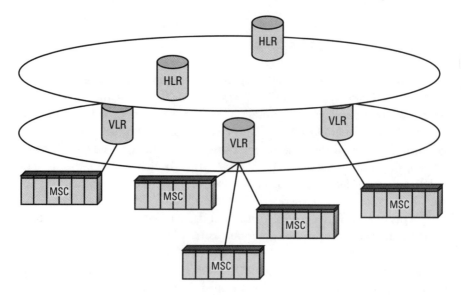

Figure 4.3 The NSS hierarchy.

Table 4.1
The Most Important Data in the HLR and the VLR

Parameter	HLR/AuC	VLR
Subscriber specific:		
IMSI	●	●
K_i	●	
TMSI		●
Service restrictions	●	
Supplementary services	●	●
MSISDN (basic)	●	●
MSISDN (other)	●	
Authentication and ciphering:		
A3	●	
A5/X (in BSS)		
A8	●	
RAND up to five triplets	●	●
SRES up to five triplets	●	●
K_c up to five triplets	●	●
CKSN		●
Subscriber location/call forwarding:		
HLR number		●
VLR number	●	
MSC number	●	●
LAI		●
IMSI detach		●
MSRN		●
LMSI	●	●
Handover number		●

The modifications of exchanges required for the provision of mobile service affect, in particular, the assignment of user channels toward the BSS, for which the MSC is responsible, and the functionality to perform and control inter-MSC handover. That defines two of the main tasks of the MSC. We have to add the interworking function (IWF), which is needed for speech and non-speech connections to external networks. The IWF is responsible for protocol conversion between CC and the ISDN user part (ISUP), as well as for rate adaptation for data services.

4.3.1 Gateway MSC

An MSC with an interface to other networks is called a gateway MSC. Figure 4.4 shows a PLMN with gateway MSCs interfacing other networks. Network operators may opt to equip all of their MSCs with gateway functionality or only a few. Any MSC that does not possess gateway functionality has to route calls to external networks via a gateway MSC.

The gateway MSC has some additional tasks during the establishment of a mobile terminating call from an external network. The call has to enter the PLMN via a gateway MSC, which queries the HLR and then forwards the call to the MSC where the called party is currently located.

4.3.2 The Relationship Between MSC and VLR

The sum of the MSC areas determines the geographic area of a PLMN. Looking at it another way, the PLMN can be considered as the total area covered by the BSSs connected to the MSCs. Since each MSC has its "own" VLR, a

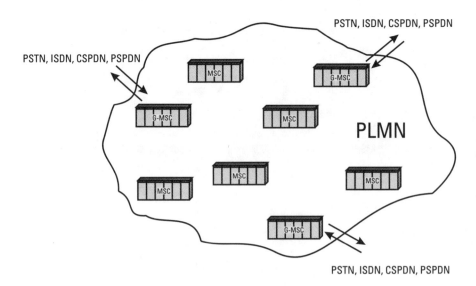

Figure 4.4 The functionality of the gateway MSC.

PLMN also could be described as the sum of all VLR areas. Note that a VLR may serve several MSCs, but one MSC always uses only one VLR. Figure 4.5 illustrates this situation.

That relationship, particularly the geographic interdependency, allows for the integration of the VLR into the MSC. All manufacturers of GSM systems selected that option, since the specification of the B-interface was not entirely available on time. In GSM Phase 2, the B-interface is no longer an open interface (as outlined above). It is expected that this trend will continue.

A network operator still has the freedom to operate additional MSCs with a remote VLR, but that is somewhat restrictive in that all the MSCs must be supplied by the same manufacturer.

4.4 Equipment Identity Register

The separation of the subscriber identity from the identifier of the MS (described in Chapter 2) also bears a potential pitfall for GSM subscribers. Because it is possible to operate any GSM MS with any valid GSM SIM, an opportunity exists for a black market in stolen equipment. To combat that, the EIR was introduced to identify, track, and bar such equipment from being used in the network.

Each GSM phone has a unique identifier, its IMEI, which cannot be altered without destroying the phone. The IMEI contains a serial number and a

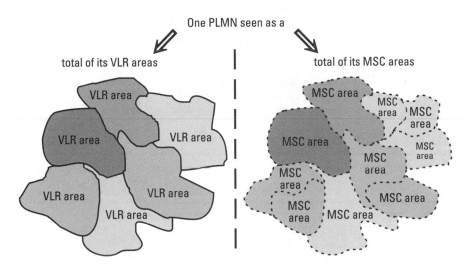

Figure 4.5 Geographic relationship between the MSC and the VLR.

type identifier. More detailed description of the structure of the IMEI is given in the Glossary.

Like the HLR or the VLR, the EIR basically consists of a database, which maintains three lists: (1) the "white list" contains all the approved types of mobile stations; (2) the "black list" contains those IMEIs known to be stolen or to be barred for technical reasons; and (3) the "gray list" allows tracing of the related mobile stations.

The prices for mobile equipment have fallen dramatically due to the great success of GSM; consequently, the theft rate is low. Several GSM operators have decided not to install the EIR or, at least, to postpone such installation for a while.

If the EIR is installed, there is no specification on when the EIR should be interrogated. The EIR may be queried at any time during call setup or location update. Chapter 12 describes this in detail.

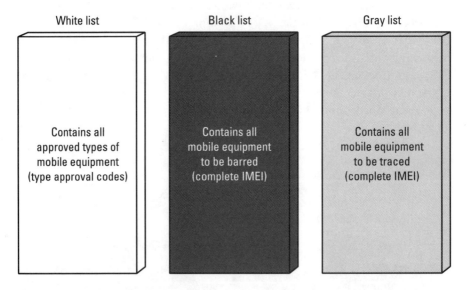

Figure 4.6 Contents of the EIR.

5

The OSI Reference Model

5.1 Reasons for Standardization

The Open System Interconnection (OSI) Reference Model was specified by ITU, in cooperation with ISO and IEC, and is documented in Recommendation X.200. Although the OSI model is applicable in many areas, it is used mostly in the area of communication between computers. Its purpose is to organize and formalize the communication method.

The basic idea of the OSI Reference Model is to separate the various parts that, in their totality, form a communications process. Separation of concerns is achieved by layering and modularization of transactions and tasks. This approach results from the following reasoning:

- The use of microprocessors in telecommunications not only allowed for the creation of new services, but at the same time increased the requirements on communications of exchanges and computers.
- When humans communicate by means of telephone or letters, they do not want to be concerned with the details of the physical transfer of the information. They just want to communicate. The same applies to the layers of the OSI Reference Model, in which each layer has a fixed role in the process of communications.
- A computer is a modular structure, which starts with the transistor on the lowest level and, with modularity built on top of modularity, ends up as a super-computer. To use such a system becomes easier when the tasks themselves can be modularized.

- Telephone systems are used for many applications. These applications became possible only because the application itself had become more independent of the process of pure data transmission. Modems and fax machines, for instance, rely on the functionality of the services of a lower layer and adapt their own functionality accordingly. The telephone network, on the other hand, is not concerned with the content or the representation of the transmitted data.
- The OSI Reference Model enables two products from different manufacturers to communicate. Many interworking problems cannot be solved simply by the interface specification shown in Figure 5.1. Each telephone set and every exchange can be regarded as network element A, B, or C. If, for example, the manufacturer of one network element implements all the functionality of a particular application and the second manufacturer implements only those functions that are necessary for a particular task, then, generally speaking, the two devices cannot be interconnected.

5.2 Layering in the OSI Reference Model

The OSI Reference Model breaks down or separates the communication process into seven independent layers. The following are the general "rules" of the OSI model:

- Two layers that lie above each other work independently. Each layer receives a service from the layer immediately below and provides a service to the layer immediately above. The lower layer does not care

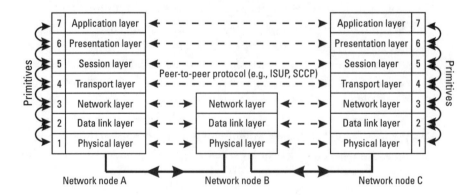

Figure 5.1 The layers and message types of the OSI Reference Model.

about the content of the received information. Consider the analogy of sending mail. The post office is not concerned about the content of a letter, which it has received to deliver. Its only concern is the address on the envelope, which it follows in order to make the delivery. This constitutes a service, which is independent of the contents of the letter.

- Each layer communicates directly only with the layers immediately below and above itself and indirectly with its peer layer at the remote end. Let us again refer to the post office analogy. Neither the sender nor the addressee is concerned about the details of the service, that is, about how the letter is routed, and neither the sender nor the addressee needs to communicate with the letter carrier. They only need to have access to a mailbox. The two parties communicate with each other by writing or reading the letter.
- If a communications process involves more than two network nodes, the intermediate network node or nodes need only provide the functionality of Layers 1 through 3. As Figure 5.1 shows, network node B is equipped only with Layers 1, 2, and 3. Layers 4 through 7 are required at the end points of a connection only. Going back to the letter delivery example, in this "communications process" all post offices and postal workers are involved only in transport and error-free delivery of the letter. All other parts of the communication process are available only at the sender and receiver sides.
- The protocols used for Layers 1 through 3 on the interface between A and B are not necessarily the same as those used on the interface between B and C. For example, Layer 2, between the BTS and the BSC in GSM, uses the LAPD protocol, while the SS7 protocol is used between the BSC and the MSC. In that case, network node B would represent the BSC.

5.3 Data Types of the OSI Reference Model

All messages exchanged between Layer N and Layer $(N-1)$ are called primitives. In practical work, except for measurements within a network element, there usually is no need to become involved in the inner workings of the primitives, it suffices to know that they exist and to have an idea of their function.

All message exchange on the same level (or layer) between two network elements, A and B, is determined by what is known as peer-to-peer protocol. Consequently, all messages that can be seen on any GSM interface between two nodes belong to the group of peer-to-peer protocols.

5.4 Information Processing in the OSI Reference Model

The tasks of the Layers 7 down to 2 mainly are to add processing information. When, for example, Layer 6 receives a primitive from Layer 7, it adds header information, which allows the "partner" layer on the other end to process the received data according to the appropriate peer-to-peer protocol. The partner layer is responsible for removing the header information. Figure 5.2 illustrates the relationship. The obvious result is the increase in the data that needs to be transmitted.

5.5 Advantages of the OSI Reference Model

The major advantage of the OSI Reference Model lies in the fact that the various layers are independent of each other. What does independence in this context mean? It means that Layer N shares a common protocol with its peer Layer N and with the layers immediately above and below it but not with any other layers. The OSI Reference Model defines only the interface between them and not the way a certain layer is implemented. Therefore, it is, for example, irrelevant to a large degree how the physical signal transmission is achieved. The transmission medium used for the data transfer may be cable, direct radio, satellite, or any other appropriate means.

That permits the design and use of modules on a general level, a statement that has to be tempered by the actual application. It certainly matters,

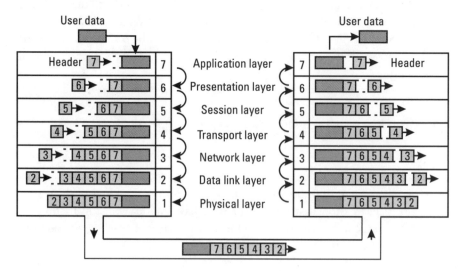

Figure 5.2 Data flow in the OSI Reference Model.

with respect to the propagation delay, whether transmission is via cable or satellite, where the propagation delays may be substantial.

5.6 The Seven Layers of the OSI Reference Model

5.6.1 Layer 1: The Physical Layer

The Physical Layer is responsible for the actual transmission of the data and the provision of the necessary facilities. The facilities can be, for example, a copper wire, a satellite connection, direct radio, or an optical fiber. The Physical Layer may include some synchronization features that do not have any significance for the higher layers, since those features are purely hardware related. Examples of such features are the clear-to-send (CTS) signal and the ready-to-send (RTS) signal of the serial interface on a computer (COM-Port).

Layer 1 does not know data types or data formats and is not able to distinguish between control data and user data. That characteristic, in particular, distinguishes Layer 1 from the other layers. The data packets received from Layer 2 are transmitted without additional verification. Each data packet consists of either a single bit or a number of bits.

With regard to the Air-interface of a GSM system, the GMSK modulation and the HF equipment in the MS and the BTS are part of Layer 1. Over the terrestrial interfaces, the PCM, including signal levels and propagation delays, is part of Layer 1.

Naturally, the implementation of the Physical Layer depends greatly on the type of interface and might change frequently. For example, between the BTS and the BSC, Layer 1 might be implemented as microwave transmission on the first section, as optical fiber on a second section, and as plain cable on a third section.

5.6.2 Layer 2: The Data Link Layer

The Data Link Layer is responsible for the packaging of the data to be transmitted. The data are combined into packets or frames and then handed to the Physical Layer for synchronous or asynchronous transmission. A widespread method for such framing is the high-level data link control (HDLC) protocol, which provides a general structure for data frames and forms, which is the basis for the SS7 protocol as well as for the LAPD protocol. The Glossary provides a description of the HDLC frame format.

The main purpose of all the tasks of Layer 2 is that of error detection and correction. Data frames are formed by introducing start/stop marks and by calculation of checksums (frame check sequence, or FCS), which can be checked for consistency by Layer 2 at the receiving side. When the receiver detects an error, it tries to correct the error or requests retransmission.

The Data Link Layer plays a vital part in protocol testing, because all data packets from Layer 3 have to be carried in a Layer 2 frame. Note that Layer 2 information is relevant only between two adjacent network nodes and that the Layer 2 protocol might change from interface to interface. For example, the Layer 2 protocol in GSM changes as the data pass on their way from the MS first at the BTS where $LAPD_m$ converts to LAPD and then again in the BSC where LAPD converts to MTP 2/SS7.

On the GSM Air-interface, Layer 2 is formed by the $LAPD_m$, together with channel coding and burst formatting. On the Abis-interface, it is LAPD, and the remaining interfaces use the MTP 2 of the SS7 protocol. Note that in $LAPD_m$, no frame check sequence is required because channel coding takes care of error detection and correction.

5.6.3 Layer 3: The Network Layer

The Network Layer prescribes the path a message has to take and who the recipient of that message is. All the information necessary to route a data packet is the responsibility of Layer 3. Layer 3 has significance only on a per-section base, as already known from Layers 1 and 2. Every network node has to analyze and possibly modify the Layer 3 information. The RR protocol between the MS, the BTS, the BSC, and the MSC belongs to Layer 3, as well as all the address information needed to route a call in an SS7 system.

The best analogy for Layer 3 information is the address information on the envelope of a letter, which has to be evaluated by every network node (post office) in its delivery path.

The equal treatment of MM, CC, and RR on the Air-interface by GSM is misleading, since MM and CC information does not belong to Layer 3. Rather, RR provides the necessary transport capability to transparently carry MM and CC information between the MS and the NSS.

5.6.4 Layer 4: The Transport Layer

Layer 4 provides the methods that guarantee the proper end-to-end ordering of message packets, before the data are handed to the higher layers (sequencing). Such handling becomes necessary when a message is partitioned into data packets. The term segmentation is used to describe the process of breaking down the information into packets. Furthermore, in contrast to the lower layers, the Transport Layer performs end-to-end data control. The Transport Layer checks the consistency of a message, when a message is composed of several pieces. The task of the Transport Layer in the OSI Reference Model is similar to that of the Data Link Layer and the Network Layer. At the time when OSI was defined, it was essential to rely on a powerful Layer 4, since Layers 2 and 3 could not handle this task alone.

The difference between Layers 2 and 3 on one side and Layer 4 on the other lies in the end-to-end application of Layer 4. While Layers 2 and 3 are relevant only on a per-interface basis, Layer 4 procedures are applied between the two end points of a connection.

A good example of a Layer 4 task is the numbering of boxes during a house move or counting them at the destination, as well as arranging them in the right order, that is, setting up the cupboard before unpacking the dishes. It is obvious that this task is not directly related to the transport or any security issue; nevertheless, the task is important for a smooth sequence of events.

5.6.5 Layer 5: The Session Layer

The Session Layer was assigned for global synchronization purposes. Both parties use the Layer 5 to coordinate the communication process between themselves. It is used in GSM between the MSC and the MS to distinguish between a mobile terminating call (MTC), a location update (LU), and a mobile originating call (MOC). Part of the synchronization is the ability to determine which information needs to be sent, when, and by whom. Another example for Layer 5 is the dialog part of the component sublayer of the transaction capabilities application part (TCAP). Two TCAP users can coordinate the type of a process, by means of the dialog part of a message and so, for example, distinguish between an LU and the activation of a supplementary service.

To come back to the example used for Layer 4: The decision about the order of packages with dishes or cupboard parts has to be made by Layer 5. Layer 4 only carries out the request.

5.6.6 Layer 6: The Presentation Layer

Generally speaking, the Presentation Layer is a means of data definition and preparation before the data are passed to the Application Layer. The Presentation Layer is able to distinguish different data types and to perform data compression and decompression. A typical example for a Layer 6 implementation is ASN.1, the Abstract Syntax Notation number 1, as defined by ITU in Recommendations X.208 and X.209.

Referring again to the analogy of the relocation, different types of boxes are necessary depending on the "data type" (i.e., cups, plates, clothing, furniture), and they need different treatment during transmission and on the receiving side (wash the dishes, set up the cupboard, etc.).

5.6.7 Layer 7: The Application Layer

The Application Layer is the interface of a specific application to the transmission medium or, in other words, to the Layers 1 through 6. Note that Layer 7 does not actually contain the application but provides an interface between the application and the communication process. Just as much as the implementation of Layer 1 depends on the physical transmission medium, so also the implementation of Layer 7 depends on the specific user.

An example best illustrates this concept, since the preceding definition is somewhat theoretical.

The president of a company does not organize a dinner party himself. That task is delegated to a third party, typically a secretary, who makes all the arrangements, including the tracking of confirmations and cancellations. The president who is not concerned with the preparation of the dinner is, in this example, the application. The third party, perhaps the president's secretary, on the other hand, does not need to be present at the dinner and does not need to know the reasons for any of the dinner speeches or to understand the reasons for inviting a certain person. The secretary has some freedom and acts independently within that area of freedom to ensure that the dinner party is well prepared and presented.

Another, more technical example is the Layer 6 of the TCAP that was specified by ITU as a general interface for all kinds of users. It is the responsibility of the users to provide Layer 6 a suitable interface, that is, a buffer. That interface or buffer is realized by Layer 7 in the according applications.

5.7 Comprehension Issues

Because of the somewhat fuzzy borders between the various layers, it is sometimes difficult to apply the OSI Reference Model to an actual problem. That is particularly true for someone who is new to the subject, and misunderstandings frequently occur.

The reason for such problems lies in the theoretical approach of the definition of the model or in the overlapping tasks of the layers. GSM adds even more complexity by switching between layers for the data transfer on the different interfaces (A, Abis, Air).

For example, when the BTS receives a Layer 2 SABM frame from the MS, it forwards that information as an EST_IND message toward the BSC, wrapped into an I-frame. The EST_IND message, clearly Layer 3 information, can be regarded as a translation of the SABM frame (Figure 5.3). This "leap" can be explained by the fact that Layers 1 through 3 are valid only on a link-by-link basis.

The remainder of this book frequently refers to the OSI Reference Model. For that reason, it is important to understand the model, its function, and the difference between the layers. The following analogy is presented to try to help give a better understanding of the "theory."

5.7.1 An Analogy: The Move to Europe

Since we all have different experiences in life and see things from different perspectives, the relationship to the OSI model is immediately emphasized during the course of this analogy, "The move to Europe."

5.7.1.1 The Moving Family as the User or Application

A family that has to move wants to have to do as little work as possible, particularly tasks like disassembly, packing, unpacking, reassembly, setting up furniture, washing dishes, cleaning rooms, and so on.

The moving family is comparable to the user or the application in the OSI model, which is outside the model. The user communicates with the moving company and defines the schedule as to when to make the move, where to move, and when the move should be finished.

5.7.1.2 The Moving Company as OSI Layers 7, 6, and 5

Let us assume, for the purpose of this analogy, that the moving company has local branches all over the world that are governed by the same business rules. The moving company has many employees: some work in the office to

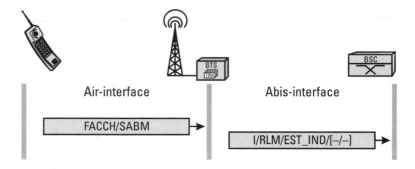

Figure 5.3 "Leap" in the layers between the air interface and the Abis interface.

coordinate the whole process, while others work onsite to do all the packing and unpacking.

The moving company has a selection of different packaging materials specifically designed for the purpose of moving household items (see *ASN.1* in the Glossary). The moving company can be likened to Layers 5, 6, and 7 of the OSI model. The employees in the office, who control the whole process, can be likened to Layer 7. The onsite employees whose function it is to separate the various household items, such as the clothes, dishes, furniture, and so on, and to pack them appropriately can be likened to part of Layer 6.

This task is similar to the processing of parameters and data in the Presentation Layer (Layer 6). The onsite employees label the boxes, according to their contents (e.g., books, clothes, dishes), which makes it easier for their counterparts in Europe to do the opposite task of unpacking. The packing and labeling procedure in Layer 5 ensures that the moving company at the destination side sets up the bookcases before unpacking the books or sets up the bureaus before unpacking the clothes. Otherwise, the books and the dishes would be unpacked before there are places to put them.

The different boxes for the various goods and the labels on the boxes are, technically speaking, peer-to-peer protocols in OSI Layers 5 and 6, which add some overhead to the process of moving household goods.

Neither the employees in the office nor those onsite deal with the actual transportation process. For that, the moving company uses the services of a transportation company.

5.7.1.3 The Transportation Company as OSI Layer 4

The transportation company is responsible for the end-to-end transportation, which is comparable to a Layer 4 task. The people who work for the transportation company count and number the boxes (error detection and segmentation) and write the destination address, based on the information they have received from the moving company (Layer 7) on the boxes. The numbering of the boxes is as requested by the moving company, that is, the employees onsite (Layer 5), who inform the transportation company as to what order (or sequence) the individual boxes have to be shipped to the destination.

Note that the numbering of the boxes creates a new peer-to-peer protocol. The transportation company does not know what the labels "bookcase," "dishes," and so on, mean in particular because they have no knowledge of the specific requirements. It is for that reason that Layer 4 translates the Layer 5 specific information into its own protocol, in this case, the numbering scheme.

When everything is done, the transportation company hands over all the boxes to the selected shipping company (Layer 3), which selects the method of physical transportation according to price and availability.

5.7.1.4 The Shipping Company as OSI Layer 3

The shipping company or its employees are equivalent to Layer 3. They are not concerned about the contents of the shipment, the numbers on the boxes, or the labels that characterize the contents. They take the boxes, process the address (routing information), and arrange for the packaging of the boxes into containers for transportation.

The long distance between the origination in America and the final destination in Europe is taken in a number of smaller steps (truck, railroad, plane, ship) and requires the reloading of the boxes into different containers. That also requires that, for each leg of the journey, addresses for the temporary, intermediate destinations have to be assigned

5.7.1.5 Truck, Railroad, Boat, and Airplane as OSI Layer 2

The various containers that have to be used, the types of which are determined mainly by the means of transport, correspond to Layer 2.

Starting at the home, a truck is used to transport the boxes to the railway station. A railroad wagon is used to transport them to the airport, harbor, and so on. The larger units correspond to the various Layer 2 protocols that are used over the link between any two nodes of a telecommunications network.

In the telecommunications environment (cargo shipping), it is Layer 2 (the means of transport) that serves the purpose of providing a secure physical transport medium for the actual data (household goods).

The checksum in a telecommunications environment corresponds to the truck driver's checklist. This illustrates the difference between what Layer 4 does for data security compared with Layer 2. While Layer 4 numbers and accounts for the boxes of one user (moving family), Layer 2 performs that task for one container that, in general, is shared by many users. That is, Layer 2 sees the process from the viewpoint of the shipping company, and Layer 4 sees it from the perspective of the user.

5.7.1.6 The Infrastructure as Layer 1

What remains is indicating the constituent parts of Layer 1. These are the roads, railroad tracks, engines, and people— everything and everyone that contributes to the physical transportation process. Just as in a move to Europe, the Physical Layer in telecommunications changes between intermediate nodes.

6

The Abis-Interface

The Abis-interface is the interface between the BTS and the BSC. It is a PCM 30 interface, like all the other terrestrial interfaces in GSM. It is specified by ITU in the G-series of recommendations. The transmission rate is 2.048 Mbps, which is partitioned into 32 channels of 64 Kbps each. The compression techniques that GSM utilizes packs up to 8 GSM traffic channels into a single 64-Kbps channel. GSM never specified the Abis-interface in every detail, as was also the case with the B-interface (the interface between the MSC and the VLR). The Abis-interface is regarded as proprietary, which leads to variations in the Layer 2 protocol between manufacturers, as well as to different channel configurations. The consequence is that, normally, a BTS from manufacturer A cannot be used with a BSC from manufacturer B.

6.1 Channel Configurations

Figure 6.1 presents two possible channel configurations of the Abis-interface. Note the fixed mapping of the air-interface traffic channels (Air0, Air1, ...) onto a time slot of the Abis-interface. This fixed mapping has the advantage that it is possible to determine which Abis time slot will be used when a particular air-interface channel is assigned.

(a)

TS	7	6	5	4	3	2	1	0	
0			FAS / NFAS						
1	Air 0	Air 1		Air 2		Air 3			TRX 1
2	Air 4	Air 5		Air 6		Air 7			
3	Air 0	Air 1		Air 2		Air 3			TRX 5
4	Air 4	Air 5		Air 6		Air 7			
5	Air 0	Air 1		Air 2		Air 3			TRX 2
6	Air 4	Air 5		Air 6		Air 7			
7	Air 0	Air 1		Air 2		Air 3			TRX 6
8	Air 4	Air 5		Air 6		Air 7			
9	Air 0	Air 1		Air 2		Air 3			TRX 3
10	Air 4	Air 5		Air 6		Air 7			
11	Air 0	Air 1		Air 2		Air 3			TRX 7
12	Air 4	Air 5		Air 6		Air 7			
13	Air 0	Air 1		Air 2		Air 3			TRX 4
14	Air 4	Air 5		Air 6		Air 7			
15	Air 0	Air 1		Air 2		Air 3			TRX 8
16	Air 4	Air 5		Air 6		Air 7			
17			not used						
18			not used						
19			partial O&M data						
20			not used						
21			O&M signaling						
22			TRX 8 signaling						
23			TRX 7 signaling						
24			TRX 6 signaling						
25			TRX 5 signaling						
26			not used						
27			TRX 4 signaling						
28			TRX 3 signaling						
29			TRX 2 signaling						
30			TRX 1 signaling						
31			not used						

(b)

TS	7	6	5	4	3	2	1	0	
0			FAS / NFAS						
1	Air 0	Air 1		Air 2		Air 3			BTS 1 / TRX 1
2	Air 4	Air 5		Air 6		Air 7			
3	Air 0	Air 1		Air 2		Air 3			BTS 3 / TRX 1
4	Air 4	Air 5		Air 6		Air 7			
5	Air 0	Air 1		Air 2		Air 3			BTS 1 / TRX 2
6	Air 4	Air 5		Air 6		Air 7			
7	Air 0	Air 1		Air 2		Air 3			BTS 3 / TRX 2
8	Air 4	Air 5		Air 6		Air 7			
9	Air 0	Air 1		Air 2		Air 3			BTS 2 / TRX 1
10	Air 4	Air 5		Air 6		Air 7			
11	Air 0	Air 1		Air 2		Air 3			BTS 4 / TRX 1
12	Air 4	Air 5		Air 6		Air 7			
13	Air 0	Air 1		Air 2		Air 3			BTS 2 / TRX 2
14	Air 4	Air 5		Air 6		Air 7			
15	Air 0	Air 1		Air 2		Air 3			BTS 4 / TRX 2
16	Air 4	Air 5		Air 6		Air 7			
17			not used						
18			partial O&M data						
19			BTS 4 / TRX 2 signaling						
20			BTS 4 / TRX 1 signaling						
21			BTS 3/ TRX 2 signaling						
22			BTS 3/ TRX 1 signaling						
23			BTS 2/ TRX 2 signaling						
24			BTS 2/ TRX 1 signaling						
25			BTS 1/ TRX 2 signaling						
26			BTS 1/ TRX 1 signaling						
27			BTS 4 / O&M signaling						
28			BTS 3 / O&M signaling						
29			BTS 2 / O&M signaling						
30			BTS 1 / O&M signaling						
31			Transmission control information						

Figure 6.1 (a) Star configuration (fullrate) and (b) serial connection (four BTSs with two TRX each).

6.2 Alternatives for Connecting the BTS to the BSC

The line resources on the Abis-interface usually are not used efficiently. The reason is that a BTS, typically, has only a few TRXs, which implies small traffic volume capability. Consequently, the line between the BTS and the BSC is used only to a fraction of its capacity. Figure 6.1(a), the star configuration, shows the case of a BTS with four TRXs, in which only 47% of the 2 Mbps actually is needed. The shaded areas mark the unused channels. When the BTS has only one TRX, that value goes down to 16%. Such waste of resources has a historical background, and it would not change much if halfrate channels were used.

When GSM specified the BTS, it defined that a BTS may have up to 16 TRXs. Two 2-Mbps interfaces are required to connect such a BTS to the BSC, because a single 2-Mbps interface is able to support only up to 10 TRXs, including O&M signaling.

Proportionally fewer resources are required on the Abis-interface when a BTS with a smaller number of TRXs is installed. The remainder cannot easily be used.

Experience has shown that the optimum for a BTS is in the range of one to four TRXs. This compromise reflects several parameters:

- Capacity. How many traffic and signaling channels does a BTS need to provide, on average and during busy hours, to avoid an overload condition?
- Available frequency range. What is the minimum distance between BTSs beyond which a given TRX frequency may be reused?

Network operators worldwide have had bad experiences, particularly with the latter point.

When digital radio was introduced, the assumption was that the impact of the disturbances, same-channel interference or neighbor channel interference, would be relatively minor. Soon after the introduction of commercial service, that assumption was found to be wrong, when more and more interference problems between BTSs appeared and degraded the quality of service. Problems with large, powerful cells were experienced, particularly in urban areas and city centers, where more and more minicells and microcells are being used.

The conclusion was to move in the direction of using more cells with fewer TRXs and smaller output power (<1W) rather than in the direction of fewer cells with more TRXs and high output power. That configuration requires a larger number of BTSs than the alternative to cover any given area. Connecting the larger number of BTSs to the BSCs, in turn, requires a larger number of links (Abis-interfaces).

Because of that trend, together with the high costs for links between the BTS and the BSC and the low efficiency when using such links, another configuration was introduced, the serial connection of BTSs.

6.2.1 BTS Connection in a Serial Configuration

In a serial configuration, the BTSs are connected in a line or a ring topology. Only one BTS, for the line topology, or two BTSs, for the ring topology, are physically connected to the BSC. Figures 6.2 and 6.3 illustrate those topologies. For the network operator, the advantage of the serial approach over the star configuration is that it saves line costs. Furthermore, the serial connection allows for more efficient use of resources, as illustrated in Figure 6.1(b). This advantage becomes particularly obvious, when colocated or sectored BTSs are used (see Section 3.1.2.3). The disadvantage, however, is that a single link failure causes the loss of the connection to a large number of BTSs

Figure 6.2 Serial connection of BTSs in a line topology. The disadvantage is that a single link failure results in total loss of connection to a number of BTSs.

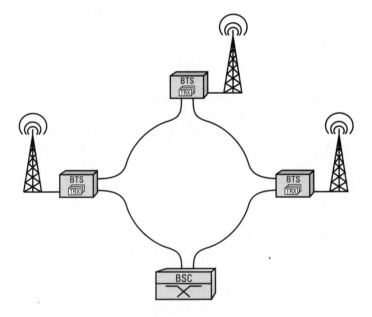

Figure 6.3 Serial connection of BTSs in a ring topology. The advantage is that a single link failure never results in total loss of connection to any BTS.

(for serial configuration). For that reason, the use of a ring configuration provides some redundancy in which the signal can always go in one of two directions, so that in the event of a link failure, it is still possible to provide an alternative connection.

6.2.2 Connection of BTSs in Star Configuration

The star configuration was the most popular when the first systems were deployed in 1991–1992. In a star configuration, every BTS has it own connection, an Abis-interface to the BSC. Figure 6.4 illustrates a star configuration with three BTSs.

6.3 Signaling on the Abis-Interface

6.3.1 OSI Protocol Stack on the Abis-Interface

The Abis-interface utilizes Layers 1 through 3 of the OSI protocol stack (Figure 6.5). Layer 1 forms the D-channel. The LAPD is in Layer 2, and Layer 3 is divided into the TRX management (TRXM), the common channel management (CCM), the radio link management (RLM), and the dedicated channel management (DCM).

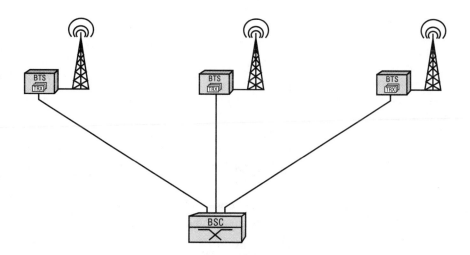

Figure 6.4 Connection of BTSs in a star configuration. The disadvantages are the high costs for links and that a single link failure always causes loss of a BTS.

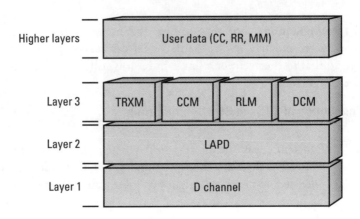

Figure 6.5 The OSI protocol stack on the Abis interface.

6.3.2 Layer 2

6.3.2.1 Link Access Protocol for D-channel

The ISDN D-channel protocol, which GSM largely has adopted, provides the basics of signaling on the Abis-interface. This link access protocol is also referred to as LAPD. The format of LAPD, as defined by ITU in Recommendations Q.920 and Q.921, is presented first before we discuss the GSM specifics. Note that GSM does not use all the functionality that ITU Q.920 and Q.921 describe. The XID frame, for example, is currently not used.

6.3.2.2 LAPD Frame

The underlying concept of the LAPD frame is the more general HDLC format, which partitions a message into an address field, a control field, a checksum, and a flag field at both ends of the message. LAPD messages in the OSI Reference Model belong to Layer 2 and are separated into three groups, according to their particular use:

- The information-frame (I-frame) group consists of only the I frame. (The unnumbered information, or UI frame, belongs to the unnumbered frame group.)
- The supervisory frame group consists of the receive-ready (RR) frame, the receive-not-ready (RNR) frame, and the reject (REJ) frame.
- The unnumbered frame group. This group comprises the set-asynchronous-balance-mode-extended (SABME) frame, the disconnected-mode (DM) frame, the UI frame, the disconnect (DISC)

frame, the unnumbered-acknowledgment (UA) frame, the frame-reject (FRMR) frame, and the exchange-identification (XID) frame.

Figures 6.6 and 6.7 illustrate the format of LAPD modulo 128 and LAPD modulo 8. The control field (defined later in the text) of the unnumbered frames is only 1 octet long (that is the case for both modulo 8 and modulo 128). The shaded area of the control field defines the message group, which is defined as follows:

- Information frame: 1st byte, bit 0 = 0
- Supervisory frames: 1st byte, bit 0 = 1, bit 1 = 0
- Unnumbered frames: 1st byte, bit 0 = 1, bit 1 = 1

Figure 6.6 and Figure 6.7 show the coding of the message type of the control field. While the group of I frames does not require any further definition, bits 2 and 3 of the first byte of a supervisory frame identify the frame type. The same task is performed by bits 2, 3, 5, 6, and 7 for the larger number of unnumbered frames.

6.3.2.3 Differences Between LAPD Modulo 128 and LAPD Modulo 8

Manufacturers have implemented LAPD differently. Some have chosen to implement LAPD modulo 8 (as shown in Figure 6.7), in which the control field consists of 8 bits, while others have chosen to implement LAPD modulo 128, which uses a 16-bit control field (as shown in Figure 6.6). Analyzing an LAPD trace file, there is no explicit possibility to distinguish between the two.

One has to rely on a consistency check, which can be performed, for example, by comparing the lengths of frames. Supervisory frames in the 8-bit version (modulo 8) are three octets long, while the ones with 16-bit-long control field (modulo 128) are four octets long. This method fails, however, for the variable-length I frames and the unnumbered frames.

On the practical side, there is only one difference between LAPD modulo 128 and LAPD modulo 8. That is the definition of the range of values for the send sequence number, N(S), and the receive sequence number, N(R). In an 8-bit-wide control field, the range for N(S) and N(R) is always between 0 and 7, while the 16-bit control field allows for values of N(S) and N(R) between 0 and 127. Hence, the two methods are referred to as LAPD modulo 8 and LAPD modulo 128, respectively.

The consequence of that is, for modulo 8, no more than eight messages may be transmitted without an acknowledgment. The difference is of little importance in GSM, since the requirement on unacknowledged frames

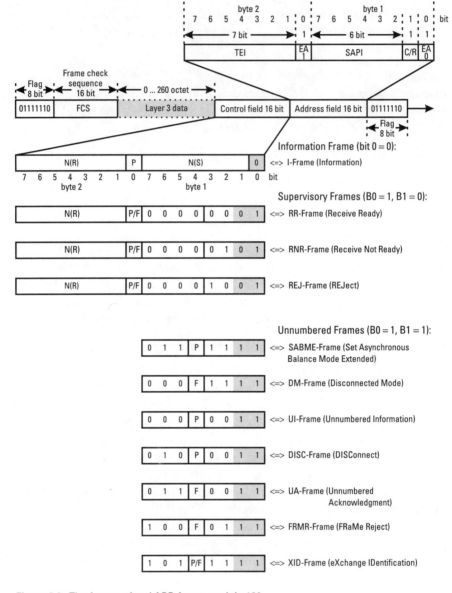

Figure 6.6 The format of an LAPD frame modulo 128.

is restricted even further by other influences. The number of unacknowledged frames for the service access point identifier (SAPI) = 0 is two, and the number of unacknowledged frames for SAPI = 62 and for SAPI = 63 is one.

The Abis-Interface 59

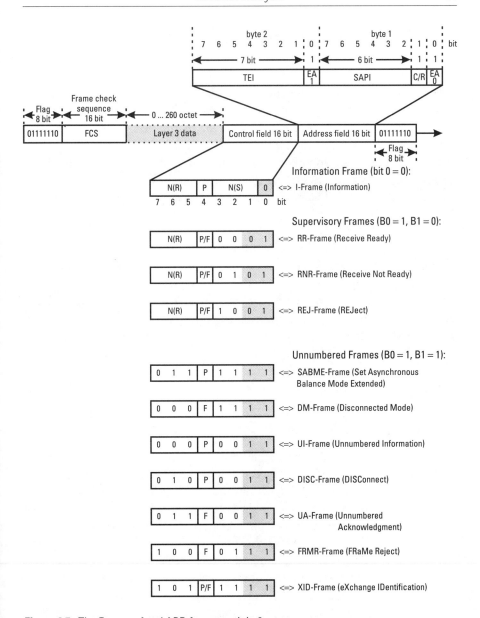

Figure 6.7 The Format of an LAPD frame modulo 8.

Nonetheless, because the modulo 128 variant is more widely used in GSM, that method is described in more detail. Furthermore, all tables and examples refer to the 16-bit variant.

6.3.2.4 Parameters of an LAPD Message

Flag

Every LAPD frame starts and ends with a flag. The flag consists of a 0-bit followed by six consecutive 1-bits and ends with a 0-bit, that is, 01111110_{bin} = $7E_{hex}$. That sequence is used as an indicator of the beginning and end of a frame. To prevent confusion, when this particular bit sequence occurs within the body of a message, some precautions need to be taken. If this pattern is part of a message, the sender has to change the sequence by inserting a 0-bit between the fifth and sixth bit. The receiver then has to remove the extra 0-bit.

Frame Check Sequence

The 16-bit long frame check sequence (FCS) is used for error detection (Figure 6.8). A checksum is calculated, using the data between the start flag and the FCS. The result is sent in the FCS field. The same operation is performed at the receiver's end, and the values of the respective FCSs are compared. The receiver will request a retransmission in the event that the calculated FCS does not match the one received.

Address Field

The parameters of the address field of a LAPD modulo 128 frame and a LAPD modulo 8 frame are described in the following paragraphs.

Service Access Point Identifier

The SAPI is a 6-bit field and defines the type of user to which a message is addressed. The functionality of the SAPI in the LAPD is similar to the function of the subsystem number (SSN) within the SCCP. SAPI is used, for instance, to determine whether a message is for O&M or if it is part of the call setup. GSM uses three different values for SAPI on the Abis-interface. Their uses are listed in Table 6.1. Note that these SAPI values are independent of those defined for the similar $LAPD_m$ standard that is used on the Air-interface. SAPI also indicates the transfer priority of a message. SAPI 62 and SAPI 63 have a

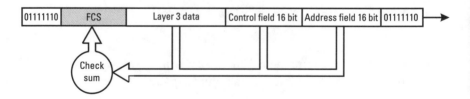

Figure 6.8 The frame check sequence.

Table 6.1
Possible Values of SAPI on the Abis-Interface

SAPI (decimal)	Priority	Meaning
0	2	Radio signaling (radio signaling link, or RSL)
62	1	O&M messages (O&M link, or OML)
63	1	Layer 2 management

higher priority for message transfer than SAPI 0. The consequence is that it is still possible, in the event of an overload situation or other problems, to exchange O&M information between the BTS and the BSC, while other information is delayed or even lost.

Terminal Endpoint Identifier

The TEI is a 7-bit field. In contrast to the SAPI, the TEI allows for distinction among several functionally identical entities. GSM uses the TEI, for example, to distinguish among the various TRXs. One TEI is assigned to each TRX. That provides the ability to distinguish between TRXs during analysis of a trace file.

Command/Response Bit

The command/response (C/R) bit determines whether a message contains a command, an answer, or an acknowledgment of a command, as illustrated in Figure 6.9 and Table 6.2. Note that the values of the C/R in a command frame are the same as the acknowledgment in the reverse direction.

As required by the ITU definition, an LAPD connection always contains a network side and a user side. When the network side sends a command, then C = 1. The user's side responds with an answer where the value of R equals 1. If a command from the user's side contains a zero value for C then the response from the network will be R = 0. There are some messages that can only be commands and others that can only be responses. In the GSM system, the BSC is defined as the network and the BTS as the user.

Extension Address Field-Bits

The address field contains one EA-bit per octet. The EA-bit of the first octet is set permanently to 0, as shown in Figures 6.6 and 6.7, which indicates that the following octet is also part of the address field. The EA-bit of the second octet is set to 1, which indicates that it is the last octet of the address field.

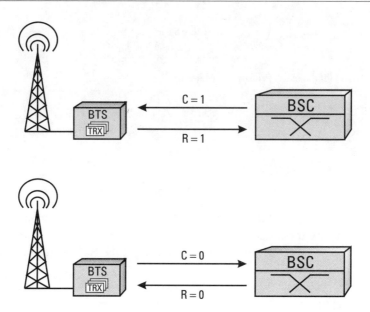

Figure 6.9 Possible values of the C/R-bit.

Table 6.2
The C/R-Bit in Command Frames and Response Frames

Frame Type	Direction	C/R?
Command frames	BSC → BTS	1
	BTS → BSC	0
Response frames	BSC → BTS	0
	BTS → BSC	1

Control Field

The length of the control field depends on the frame type and is either 8 or 16 bits long. It contains the following information.

Polling Bit (P-Bit), Final Bit (F-bit), and P/F-Bit For frame types that can be used only as commands, the corresponding bit is the P-bit. In frames that can be used only as responses, the corresponding bit is the F-bit. In frame types that can be used as both commands and responses, all variants are possible. The P-bit informs the receiver of a command message that the sender expects an answer, even if the message type normally would not require an acknowledgment. Real polling on the Abis-interface is used only when BSC and BTS are in an idle state and need to test the connection periodically (i.e., the exchange of RR frames).

When a command frame is received where the P-bit is set to 1, the answer frame needs to be returned with the F-bit set to 1. LAPD allows for the acknowledgment of an I frame, where the P-bit is set to 0, with either an I frame or a supervisory frame. I frames, however, where the P-bit is set to 1, have to be acknowledged immediately with a supervisory frame. The P-bit of UI frames is always set to 0. That is why a UI frame, although a command per definition, does not require an acknowledgment. Example 6.3 describes polling for an RSL.

Send Sequence Number and Receive Sequence Number The N(S) and the N(R) serve the purpose of acknowledging the transfer and the receipt of I frames. The method of counting can be modulo 8 or modulo 128. In the case of modulo 8, three bits are used for the counter, allowing for values of frame numbers between 0 and 7. Seven bits are used for the counter in the case of modulo 128, allowing for values between 0 and 127. On the Air-interface (LAPD$_m$), only modulo 8 is used, whereas both variants are used on the Abis-interface. The functionality as such is independent of the value range of the counters. When one side (BSC or BTS) sends an I frame, the counter N(S) on the sender side is incremented by 1. Note that the value of N(S) in the just sent I frame still has the old value, that is, the increment occurs only after the frame is sent.

When an I frame reaches the receiver, it is checked to see if the received values of N(S) and N(R) match those the receiver has stored. The value for N(S) for the received I frame has to match the actual value of N(R) on the receiver side. If the frame also is without errors (FCS), the receiver increments the value for N(R) and sends that new value in an RR frame back to the sender. The sender expects acknowledgment within a specified time frame. If that time period expires without the acknowledgment, the I frame is sent again. Note that according to Specifications Q.920 and Q.921, an acknowledgment does not have to be given by a supervisory frame but also can be given by an I frame. Consequently, the sending of an RR frame is not necessary if the receiver has to send an I frame, too. However, GSM does not make use of that option. Every I frame gets acknowledged with an RR frame. Until the acknowledgment is received, the sender has to buffer an I frame. The following example illustrates this strategy.

Function of N(S) and N(R) The BTS sends an I frame and increments its counter N(S). The BSC receives the I frame, increments counter N(R), and sends an RR frame with a new value of N(R) back to the BTS. The BTS does not need to continue to buffer the I frame after it receives the acknowledgment from the BSC.

Next the BSC sends an I frame to the BTS and increments its counter N(S) to 1. Again, note that the values of N(S) and N(R) in the transmitted I frames correspond inversely to the ones stored internally in the BTS. The BTS then checks for consistency of the information and increments, if everything is right, its counter N(R) and responds to the BSC with an RR frame with the new value of N(R). This procedure is illustrated in Figure 6.10.

RR frames need to be exchanged between BTS and BSC within certain time intervals during the so-called idle case, when no data are being transported. The values of N(S) and N(R) are not changed during that process, which is called polling. However, they have to correspond inversely to each other.

This applies to both LAPD modulo 8 and to LAPD modulo 128.

Frame Type The control field identifies, among other things, the frame type. Table 6.3 lists which values (in hexadecimal) the control field of an LAPD frame modulo 128 can assume in a trace file. Digits marked with an X indicate a "don't care" condition, that is, the value of the digit is irrelevant in identifying the frame type.

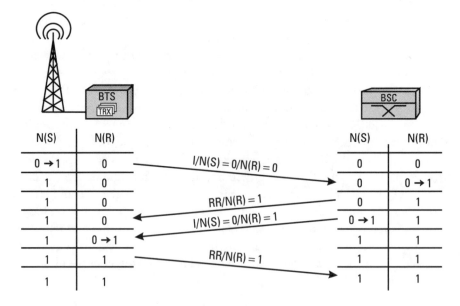

Figure 6.10 Function of the counters N(S) and N(R).

Table 6.3
Frame Types of the Abis-Interface

Name	Command-Frame?	Response-Frame?	Possible Values of the Control Field (in Hex)
I-frame group			
I	Yes	No	(X0 XX), (X2 XX), (X4 XX), (X6 XX), (X8 XX) if 1st byte is even it is an I-Frame
Supervisory frame group			
RR	Yes	Yes	(01 XX)
RNR	Yes	Yes	(05 XX)
REJ	Yes	Yes	(09 XX)
Unnumbered frame group			
FRMR	No	Yes	(87), (97)
DISC	Yes	No	(53) because P-bit is always 1
UI	Yes	No	(03) because P-bit is always 0
DM	No	Yes	(0F), (1F)
SABME	Yes	No	(7F) because P-bit is always 1
UA	No	Yes	(73) because P-bit is always 1
XID	Yes	Yes	(AF), (BF) (not used in GSM)

Detection of Frame Type of LAPD Frames The LAPD messages in Figure 6.11 have been recorded on the Abis-interface by means of a low-level protocol analyzer. For our imaginary analysis, complete decoding is not required; only the message types are identified.

It is known that the first three octets in line number 0010 of the UBFD (user buffer = trace file specific) always contain the address and the control field of the respective LAPD message.

Note how difficult it would be without this knowledge to identify the relevant information in a given trace file. If one encounters such a situation, the best way to proceed is to look for supposedly included fields like the address and control fields within an LAPD message by using a regular editor (see Figure 6.10).

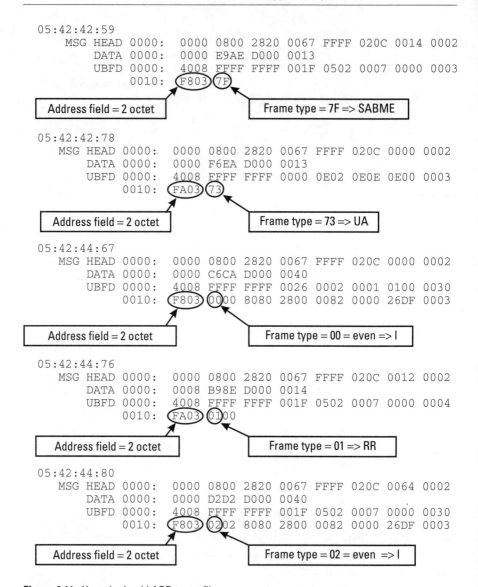

Figure 6.11 Hexadecimal LAPD trace file.

Tasks of Various Frame Types

I Frame The I frame (Figure 6.12) is used to transfer Layer 3 information. It is always a command, irrespective of its direction. The error-free reception of this frame has to be acknowledged by the recipient with an RR frame.

```
                                              Bit
 15  14  13  12  11  10  9  8  7  6  5  4  3  2  1  0
 |       N(R)          | P |      N(R)         | 0 |
```

Figure 6.12 Control field of an I frame (modulo 128).

Otherwise, an RNR frame or an REJ frame is sent, because the frame could not be processed due to some error or overload condition and thereby requests retransmission of the I frame. I frames contain both an N(S) and an N(R).

RR Frame An RR frame (Figure 6.13) acknowledges that an I frame has been received. It also is used for the polling between the BTS and the BSC.

During idle phases (no I-frame transmission), RR frames are exchanged between the BSC and the BTS with a periodicity based on the value of timer T203 (the default value of T203 is 10 seconds). It may be assumed that Layer 2 of a connection is working fine when polling of RR frames can be seen on the Abis-interface. No conclusion, however, can be drawn as to the state of the Layer 3 connection.

RNR Frame The RNR frame (Figure 6.14) is used to signal that no more I frames can be accepted. This situation may arise when too many unprocessed I frames are stored in the input buffer, so that no space is available for more I frames. In this situation, an RNR frame is sent to the remote end.

The RNR frame requests a halt to the transmission of I frames and requires the transmitter to wait for an RR frame before transmission can be resumed.

This frequently results in an overload situation on the sender side because data for transmission quickly backs up, which, in turn, results in the sender also

```
                                                  Bit
 15  14  13  12  11  10  9  8   7  6  5  4  3  2  1  0
 |        N(R)         |P/F| 0  0  0  0 | 0  0  0  1 |
```

Figure 6.13 Control field of an RR frame (modulo 128).

```
                                                  Bit
 15  14  13  12  11  10  9  8   7  6  5  4  3  2  1  0
 |        N(R)         |P/F| 0  0  0  0 | 0  1  0  1 |
```

Figure 6.14 Control field of an RNR frame (modulo 128).

sending an RNR frame. The value of N(R) in the RNR message is that it indicates which I frame was last received correctly.

REJ Frame In contrast to the RNR frame, which is used to signal an overload situation and hence to request the temporary halt to transmission, the REJ frame (Figure 6.15) is used to indicate a transmission error condition that has been detected by analysis of the FCS. The REJ frame contains a value for N(R), which indicates the first I frame that has to be repeated.

An REJ frame is also used to indicate that I frames with a wrong value for N(S) or N(R) were received. That requests the retransmission of all I frames with a value of N(R) and higher.

SABME Frame SABME frames (Figure 6.16) are sent when no Layer 2 connection has been established.

DM Frame The transmitting side uses a DM frame (Figure 6.17) to indicate that it can no longer maintain the Layer 2 connection.

A DM frame indicates that the sender will immediately tear down the Layer 2 connection without waiting for an acknowledgment from the receiver. The DM frame is used to take a connection out of service, as is the case with the DISC frame, but without waiting for or expecting an acknowledgment.

UI Frame Unlike an I frame, a UI frame (Figure 6.18) contains neither a send sequence number nor a receive sequence number. Another difference is that the

15	14	13	12	11	10	9	8	7	6	5	4	3	2	1	Bit 0
			N(R)				P/F	0	0	0	0	1	0	0	1

Figure 6.15 Control field of an REJ frame (modulo 128).

7	6	5	4	3	2	1	Bit 0
0	1	1	P	1	1	1	1

Figure 6.16 Control field of a SABME frame.

7	6	5	4	3	2	1	Bit 0
0	0	0	F	1	1	1	1

Figure 6.17 Control field of a DM frame.

```
          Bit
7 6 5 4 3 2 1 0
0 0 0 P 0 0 1 1
```

Figure 6.18 Control field of a UI frame.

content of a UI frame does not require an acknowledgment (P-bit = 0). UI frames are used on the Abis-interface to convey MEAS_RES messages.

DISC Frame   The DISC frame (Figure 6.19) is used to take a Layer 2 connection out of service. The transmitter informs its peer that it intends to tear down the Layer 2 connection. In contrast to the DM frame, the DISC frame is used for regular maintenance tasks, and an acknowledgment (UA) is expected.

UA Frame   The UA frame (Figure 6.20) is used to answer a SABME frame or a DISC frame. It acknowledges a Layer 2 connection being brought into service as well as one taken out of service.

FRMR Frame   The FRMR frame (Figure 6.21) indicates that a received message was garbled, wrong, or unexpected (protocol error). That is different from the REJ frame, which indicates to the peer entity that I frames have to be repeated starting at N(R).

This kind of error cannot be corrected by retransmission of a frame. The problem usually occurs when there are errors with the transmission between

```
          Bit
7 6 5 4 3 2 1 0
0 1 0 P 0 0 1 1
```

Figure 6.19 Control field of a DISC frame.

```
          Bit
7 6 5 4 3 2 1 0
0 1 1 F 0 0 1 1
```

Figure 6.20 Control field of a UA frame.

```
          Bit
7 6 5 4 3 2 1 0
1 0 0 F 0 1 1 1
```

Figure 6.21 Control field of an FRMR frame.

the BTS and the BSC. The FRMR frame may be sent as an answer to any frame, and its use is not restricted to being a response to faulty I frames.

To allow for proper identification of the faulty frame, the header and as much content of the faulty frame as possible are sent back to the peer entity within the FRMR frame.

Problems in the transmission area or at the opposite end of the connection can be assumed when FRMR frames are detected.

XID Frame   Although the XID frame is not part of the Abis-interface, according to GSM 08.56, it will be described here.

The XID frame (Figure 6.22) is used to synchronize the various transmission parameters between the user and the network. It coordinates the various timers when the Layer 2 connection is brought into service. It also determines the number of unacknowledged I frames that have to be stored.

Example: Decoding and Analysis of Layer 2 Messages   To decode and analyze Layer 2 messages when observing a protocol analyzer, one needs to be able to interpret the displayed measurement results. This example describes such decoding work by analyzing the following trace file. Note that the channels are mostly in idle state because no data have to be transferred between the TRX and the BSC. RR frames are exchanged periodically between the TRX and the BSC according to the value of the timer T203. Sometimes ghost CHAN_REQ messages are decoded by the BTS receiver, created by the unavoidable electromagnetic noise in the environment. Unfortunately, that false CHAN_REQ decoding may put an additional signaling load on the system. Figure 6.23 presents a short section of a trace file that was captured with the SIEMENS K1103 protocol analyzer.

- Note the following details, which can be retrieved from a rather uninteresting trace file without any error messages.
- Timer T203, in this case, is set to 3.4 seconds. The time can be derived from the time difference between two consecutive RR frames from the same TRX.

7	6	5	4	3	2	1	Bit 0
1	0	1	P/F	1	1	1	1

Figure 6.22   Control field of an XID frame.

The Abis-Interface 71

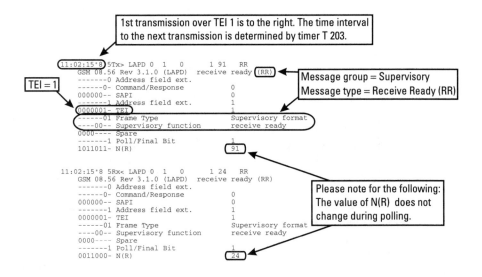

Figure 6.23(a) Tracefile as taken from a K1103.

- The messages from left to right are sent by the TRX, while the messages from right to left are sent by the BSC. That can be deduced from the CHAN_RQD at the end of the trace file, a message that can come only from the BTS and in this case is sent from left to right.
- The P/F-bit is set to 1 during polling to request the distant end to answer. For "normal" data exchange, the P/F-bit is set to 0.

In this example, the TRX polls the BSC.

6.3.3 Layer 3

Layer 3 information within I frames and UI frames follows the Layer 2 header. Because of differences in format, it is particularly important during protocol testing to distinguish between Layer 3 information for administrative tasks (SAPI 62 and 63) and Layer 3 information for connection setup and release (SAPI 0).

SAPI 0 is allocated to the RSL and carries user signaling, that is, all messages for connection setup and release. On the Abis-interface, messages for SMS and supplementary services (SS) also are dedicated to SAPI 0, which differs from the handling on the Air-interface.

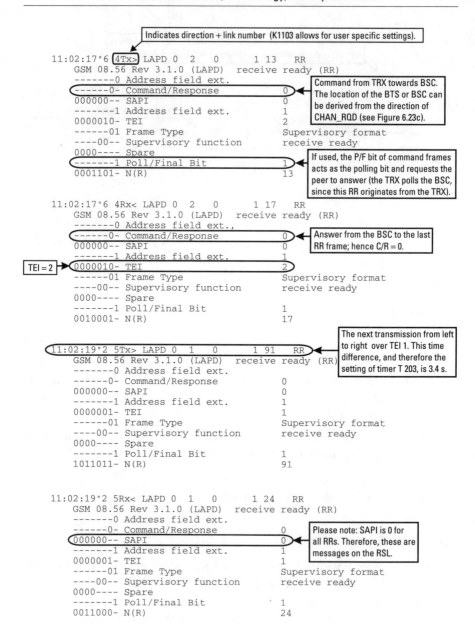

Figure 6.23(b) Tracefile as taken from a K1103.

Administrative data, on the other hand, are assigned to SAPI 62 and 63. Administrative tasks are control commands from the BSC (OMC) to the BTS,

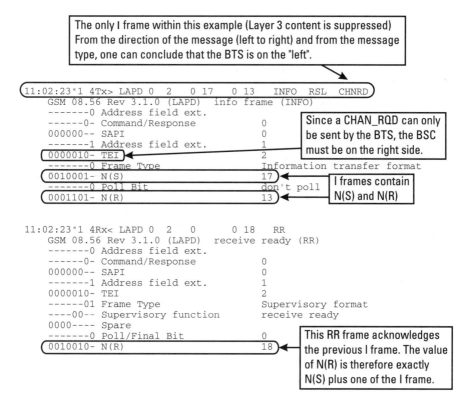

Figure 6.23(c) Tracefile as taken from a K1103.

as well as complete software packages and files, which the BSC (OMC) sends to the BTS.

6.3.3.1 Layer 3 on RSL (SAPI 0)

Figure 6.24 illustrates the format of Layer 3 on the RSL. The parameters are described in more detail. Note at this point how Layer 3 is embedded in Layer 2.

Message Discriminator and the T-Bit

The message discriminator classifies all the messages defined in Layer 3 of the Abis-interface into groups or classes (see Figure 6.5). Together, the groups form Layer 3 on the Abis-interface. The purpose of the T-bit indicates whether the BTS should process an incoming message (e.g., MEAS_RES) or if the message should be transparent to the BTS (T = 1). The distinction applies to both the uplink and the downlink.

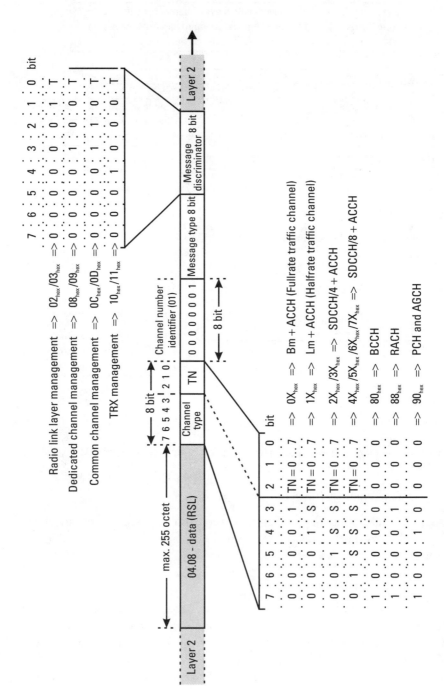

Figure 6.24 Formatting of Layer 3 signaling on the RSL.

Figure 6.24 shows the hexadecimal values the message discriminator can take, depending on the T-bit and how those values translate into the different message classes.

The classes organize messages according to their use:

- RLM. This group contains all the messages necessary for the control of a Layer 2 connection between the MS and the BTS. That includes connection setup and release, as well as the reporting of Layer 2 problems on the Air-interface to the BSC. The DATA_REQ and the DATA_IND messages, which are used for the transparent signaling data transport between MSC and MS, also belong to this group.
- CCM and TRXM. All messages that carry common control channel (CCCH) signaling data to and from the Air-interface are assigned to the CCM. That includes the transfer of cell broadcast information to the BTS. Messages used for TRXM also belong to this group.
- DCM. All messages that are used to control Layer 1 of the Air-interface belong to DCM.

Message Types

Tables 6.4, 6.5, and 6.6 list all the messages that are defined on the Abis-interface. The letters in message names that appear in uppercase form the message mnemonics used in the context and the protocol presentations.

Channel Number

The channel number is a parameter that identifies the channel type, the time slot, and the subchannel that are used for a connection on the Air-interface. Note that the channel number only indirectly corresponds to the terrestrial channel used on the Abis-interface. This parameter consists of an element identifier, which is hard coded to 01_{hex} plus the actual information, which is formatted as shown in Figure 6.24 and in Table 6.7.

- The S-bits specify the subchannel (if required) and can take a value in the range from 0 to 7.
- The X-bits identify the time slot (not the frequency) on the Air-interface and can take a value in the range from 0 to 7.
- Table 6.7 shows that it is easy to derive the channel type from the hexadecimal representation. For example, channels 08, 09, 0A, ..., are all fullrate traffic channels, because the first digit is set to 0, which applies only to the fullrate traffic channel.

Table 6.4
Messages for RLM

ID (Hex)	Name	Direction	Explanation
01	DATA REQest	BSC → BTS	Transport container for the transparent transfer of BSSAP data from the NSS to the mobile station.
02	DATA INDication	BTS → BSC	Transport container for the transparent transfer of BSSAP data from the MS to the NSS.
03	ERROR INDication	BTS → BSC	Informs the BSC of a problem in Layer 2 of the air interface (e.g., acknowledgments are missing, MS sends $LAPD_m$ frames with a wrong address or control field, radio link failure/Layer 2). Note that not every ERR_IND message in a protocol trace means a dropped call.
04	ESTablish REQest	BSC → BTS	Request for the BTS to establish a Layer 2 connection on the Air-interface. The BTS subsequently sends an $LAPD_m$ SABM frame to the mobile station.
05	ESTablish CONFirm	BTS → BSC	Answer to EST_REQ. Message sent to the BSC after the BTS receives an $LAPD_m$ UA frame from the mobile station.
06	ESTablish INDication	BTS → BSC	Response from the BTS on receiving an $LAPD_m$ SABM frame from the mobile station. *Note:* A SABM on the Air-interface may contain Layer 3 data (Example: LOC_UPD_REQ). In that case, EST_IND contains Layer 3 data.
07	RELease REQest	BSC → BTS	Request to a BTS to release an existing Layer 2 connection on the air interface. After the BTS receives this, it sends an $LAPD_m$ DISC frame to the mobile station.
08	RELease CONFirm	BTS → BSC	Answer to REL_REQ. Message sent to the BSC after the BTS has received an LAPDm UA answer frame (for an LAPDm DISC frame) from the mobile station.
09	RELease INDication	BTS → BSC	Response from the BTS when receiving an $LAPD_m$ DISC frame from the mobile station. In this case, the BTS releases the Layer 2 connection without waiting for a response from the BSC.
0A	UNIT DATA REQest	BSC → BTS	Transport frame for messages sent in $LAPD_m$ UI frames over the Air-interface (downlink).
0B	UNIT DATA INDication	BTS → BSC	Transport frame for messages received by the BTS as $LAPD_m$ UI frames (uplink).

Table 6.5
Messages for CCM and TRXM

ID (Hex)	Name	Direction	Explanation
11	BCCH INFOrmation	BSC → BTS	Transport frame for SYS_INFO messages where the BTS permanently transmits SYS_INFO 1–4 in the BCCH time slot 0.
12	CCCH LOAD INDication	BTS → BSC	Informs the BSC about the traffic load on the common control channels of the Air-Interface. The frequency of transmission of CCCH_LOAD_IND may be adjusted by the OMC.
13	CHANnel ReQuireD	BTS → BSC	Message sent by the BTS after a CHAN_REQ message is received from the mobile station. The CHAN_RQD message actually contains more information than CHAN_REQ, in particular the TA parameter, which is determined by the BTS.
14	DELETE INDication	BTS → BSC	Infrequent message caused by an overload situation on the Air-interface when a common control channel could not be sent.
15	PAGing CoMmanD	BSC → BTS	Response of the BSC on receiving a PAGING command from the MSC. Contains IMSI and/or TMSI and the paging group of the called mobile station.
16	IMMediate ASSign CoMmanD	BSC → BTS	Contains all information for assignment of a SDCCH on the Air-Interface. Used by the BSC as a response on receiving a correct CHAN_RQD.
17	SMS BroadCast REQest	BSC → BTS	Transfers cell broadcast service (CBS) messages to the BTS, which the cell broadcast center (CBC) sends to all mobile stations within a given area. The SMS_BC_REQ message can transfer only 22 octets of data (plus one octet header), as opposed to the SMS_BC_CMD message. As a result, it requires four SMS_BC_REQ messages to transport one complete CBS message with a length of 88 octets.
19	RF RESource INDication	BTS → BSC	The BTS uses this message to periodically inform the BSC about quality and quantity of the available resources on the air interface. The information on quality is derived from the idle-channel measurements of the TRX receivers. It enables the BSC to refrain from assigning channels with lower quality.
1A	SACCH FILLing	BSC → BTS	Message sent to the BTS, together with the BCCH-INFO, when a TRX is put into service. The SACCH_FILL message informs the TRXs as to which data of an active connection should be transmitted in the SYS_INFO 5-6.

Table 6.5 (continued)

ID (Hex)	Name	Direction	Explanation
1B	OVERLOAD	BTS → BSC	Message to prevent an overload on the Air-Interface or the TRX. After receiving this message, the BSC reduces the rate of transmission of new messages. The rate is further reduced when more OVERLOAD messages are received from the BTS. If no more OVERLOAD messages are received within the time frame, defined by timer T2, the transfer rate is increased gradually.
1C	ERROR REPORT	BTS → BSC	The ERROR_REPORT message is used when a protocol error has been detected by the TRX and no other response message exists. The ERROR_REPORT message contains, among others, the message that could not be processed. Reason for such errors could be (1) an undefined message type or message discriminator (e.g., caused by bit errors or protocol errors on the Abis-interface) or (2) that the TRX is unable to activate ciphering requested by the BSC.
1D	SMS BroadCast CoMmanD	BTS → BSC	The SMS_BC_CMD message is a new GSM Phase 2 message that is used for transfer of CB/SMSCB messages. Its task corresponds to the SMS_BC_REQ message but with the proviso that the SMS_BC_CMD message is capable of carrying all 88 octets of a CBS message at one time.

Table 6.6
Messages for DCM

ID (Hex)	Name	Direction	Explanation
21	CHANnel ACTivation	BSC → BTS	Message to reserve and activate channels on the Air-interface. It is sent before seizure and contains an accurate description of the requested channel (halfrate/fullrate, DTX on/off, channel type, etc.). The BTS needs this information to activate the transcoders (TRAU). An example for this process can be found in Chapter 3.
22	CHANnel ACTivation ACKnowledge	BTS → BSC	The BTS acknowledges with this message the reception of a CHAN_ACT message and activation of the requested channel. *Note:* A reception of a CHAN_ACT_ACK message does not necessarily indicate that a channel activation was performed without error.

Table 6.6 (continued)

ID (Hex)	Name	Direction	Explanation
23	CHANnel ACTivation Negative ACKnowledge	BTS → BSC	Negative answer to the CHAN_ACT message. The BTS could not provide the channel requested by the BSC. Possible reasons include overload and function not available.
24	CONNection FAILure	BTS → BSC	An important message for error analysis on the Abis-interface. CONN_FAIL is sent by the BTS in case of Layer 1 problems on the air interface (e.g., radio link failure/Layer 1).
25	DEACTivate SACCH	BSC → BTS	Request sent to the BTS to stop transmission over the slow associated control channel (SACCH). The DE-ACT_SACCH message is part of the release procedure.
26	ENCRyption CoMmanD	BSC → BTS	Activation of ciphering on the Air-interface. The ENCR_CMD message informs the BTS of which algorithm A5/X is to use and contains the complete CIPH_MOD_CMD message destined for the mobile station.
27	HANDover DETect	BTS → BSC	HND_DET is used during handover (not for intra-BTS and intra-BSC). After the target cell has received the HND_ACC message, it calculates the distance to the MS (TA) and sends the result in the HND_DET message to the BSC. In addition, the purpose of HND_DET is to inform the MSC about a successful handover as early as possible, to allow for a faster switching of the user channel on the A-interface, even before the HND_COM message is received (reduction of dead time during handover).
28	MEASurement RESult	BTS → BSC	Contains the mutual measurement results of the BTS and the MS. The MEAS_REP, which comes from the MS, is incorporated into a MEAS_RES message.
29	MODE MODify REQuest	BSC → BTS	GSM allows the channel mode on the air interface to be changed even during an active connection. If certain characteristics need to be adjusted, a MODE_MOD_REQ message is sent to the BTS that tells the BTS the new settings (e.g., switching between data and speech, DTX on or off).
2A	MODE MODify ACKnowledge	BTS → BSC	Confirmation for MODE_MOD_REQ. The BTS has received and processed the message, that is, the new transmission parameters have been adopted.
2B	MODE MODify Negative ACKnowledge	BTS → BSC	Negative response for a MODE_MOD_REQ message. The BTS is unable to provide the requested channel characteristics.

Table 6.6 (continued)

ID (Hex)	Name	Direction	Explanation
2C	PHYsical CONTEXT REQuest	BSC → BTS	Is used by the BSC to query the BTS about the latest values for the distance between the MS and the BTS, the power level, or channel type.
2D	PHYsical CONTEXT CONFirm	BTS → BSC	Answer to PHY_CONTEXT_REQ. The BTS provides the requested information to the BSC.
2E	RF CHANnel RELease	BSC → BTS	The RF_CHAN_REL message is sent to the BTS after the release of the Layer 2 connection on the air interface, to release the physical channel (Layer 1).
2F	MS POWER CONTROL	BSC → BTS	The MS_POWER_CON message is used by the BSC to adjust the output power of an MS according to the current radio conditions. The value range depends on the standard (GSM, DCS 1800, PCS 1900) and on the power class of the MS and ranges from 20 to 30 dB. Adjustments can be done in steps of 2 dB.
30	BS POWER CONTROL	BSC → BTS	The BS_POWER_CON message is used by the BSC to adjust the output power of the BTS. The value range is 30 dB and is independent of the type of standard (GSM, DCS 1800, PCS 1900). Adjustments can be done in steps of 2 dB.
31	PREPROCess CONFIGure	BSC → BTS	The PREPROC_CONF message is used only when the BTS is in charge to preprocess its own measurements and those received from the MS. In that case, not all measurement results are forwarded to the BSC, but the BTS preprocesses all data and periodically sends only a PREPRO_MEAS_REAS message to the BSC. The details of the preprocessing are manufacturer-dependent and may be configured by the network operator. The PREPROC_CONF message is necessary to coordinate the details of preprocessing between BSC and BTS.
32	PREPROcessed MEASurement RESult	BTS → BSC	See PREPROC_CONF.
33	RF CHannel RELease ACKnowledge	BTS → BSC	The RF_CH_REL_ACK message acknowledges that an RF_CHAN_REL message was received and processed. The BTS has released a previously occupied physical channel.
34	SACCH INFO MODIFY	BSC → BTS	The SACCH_INFO_MODIFY message is used to modify the SYS_INFOS 5 and 6, which are sent during an active connection. The BTS receives the original information about their content in the SACCH_FILL message.

Table 6.7
Coding of the Channel Number

7	6	5	4	3	2	1	0	Possible Values (in Hex)	Channel Type
0	0	0	0	1	X	X	X	0X	TCH (fullrate)
0	0	0	1	S	X	X	X	1X	TCH (halfrate)
0	0	1	S	S	X	X	X	2X, 3X	SDCCH/4
0	1	S	S	S	X	X	X	4X, 5X, 6X, 7X	SDCCH/8
1	0	0	0	0	0	0	0	80	BCCH
1	0	0	0	1	0	0	0	88	Uplink CCCH (RACH)
1	0	0	1	0	0	0	0	90	Downlink CCCH (PCH, AGCH)

Figure 6.25 shows an example of an RF_CHAN_REL message.

Decoding of Abis-Interface Messages (Layers 2 and 3)

Consider the following sequence of hexadecimal numbers taken from a trace on the Abis-interface. As usual, there was no protocol analyzer to decode from hexadecimal to mnemonics. Therefore, the analysis needs to be done "manually."

00 03 E2 D3 0C 13 01 88 13 E6 9D AB 11 00

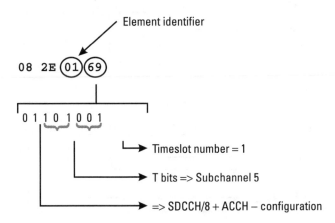

Figure 6.25 The RF_CHAN_REL message in hexadecimal format.

Decoding of Layer 2 Figures 6.6 and 6.7 are necessary to decode the Layer 2 data, illustrated in Figure 6.26. Note carefully that flags and the FCS are not contained in the preceding sequence. That is valid for most messages you will have to decode in your daily work.

Note the shaded bit in the first octet of the control field. That bit identifies the whole frame as an I frame (modulo 128).

The following information can be derived from the address and control fields:

- SAPI = 0 indicates that this is an RSL message.
- C/R = 0 and frame type = I frame indicate that this is a command from the BTS to the BSC.
- P = 1 indicates that the sender of the message expects an acknowledgment (RR frame).
- As indicated in Table 6.3, the frame is readily identified as an I frame by noticing that the first byte of the control field is E2, that is, an even number. That frequently is sufficient to make an identification of the frame type. Other examples are 01 for an RR frame or 03 for a UI frame.

Decoding of Layer 3 Decoding of Layer 3 data (Figure 6.27) on the Abis-interface is performed with the aid of GSM Recommendations 04.08 and 08.58 and with reference to Figure 6.24.

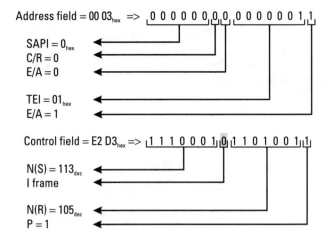

Figure 6.26 The decoding of Layer 2.

00 03 E2 D3	0C 13 01 88 13 E6 9D AB 11 00
Layer 2	Layer 3

Figure 6.27 Layer 2 and Layer 3 of the message.

Layer 3 information follows the address and the control fields, as illustrated in Figures 6.6, 6.7, and 6.24.

Figure 6.24 shows the format of Layer 3 data on the Abis-interface. It begins with one octet for the message discriminator (indicating the message group), one octet for the message type, then two bytes to specify the channel number. The message type determines the interpretation of the remainder of the message. Figure 6.27 would be interpreted as is shown in Figure 6.28 (all data are in hexadecimal).

Therefore, the remaining octets, 13 E6 9D AB 11 00, belong to the connection request of an MS. The decoding is illustrated in Figure 6.29 (taken from GSM 08.58).

6.3.3.2 Layer 3 on the Operation and Maintenance Link (SAPI 62)

Before describing the data format of SAPI 62, it should be mentioned that the differences between the various manufacturers become particularly obvious here. It is possible that in certain cases, the presented formats differ from your actual recordings.

Different data formats have to be used on the OML, the connection between BSC and the O&M unit in the BTS, from those used for connection setup and release. Transfer of operations software to the BTS and forwarding of maintenance commands fall into that category. In particular, for software transfer, a greater amount of data has to be transferred from the BSC to the BTS, which requires the segmentation and sequencing of the messages (refer to OSI Layer 4).

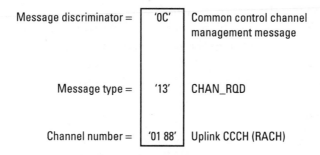

Figure 6.28 Identification of the Layer 3 header.

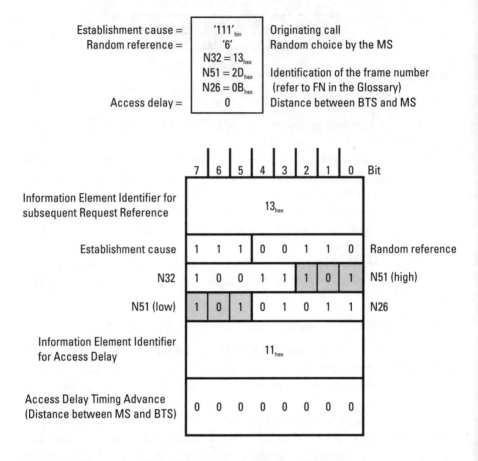

Figure 6.29 Format of a CHAN_RQD message (taken from GSM 08.58).

Although the format depends on the manufacturer, GSM still provides some guidelines in Recommendations GSM 08.59 and GSM 12.21. But because GSM 12.21 only recently has become available, many manufacturers still use their own proprietary protocols. Within those protocols, the manufacturers also implement the higher layers of the OSI model, which are missing on SAPI 0 (RSL).

Figure 6.30 shows the general format for data transfer on the OML.

Frame Structure

The Layer 2 differences for the data destined for SAPI 0, 62, or 63 were demonstrated in the discussion of Layer 2 and are not shown in Figure 6.30 (see Figures 6.6 and 6.7). The first octet of the O&M Layer 3 is an identifier, which

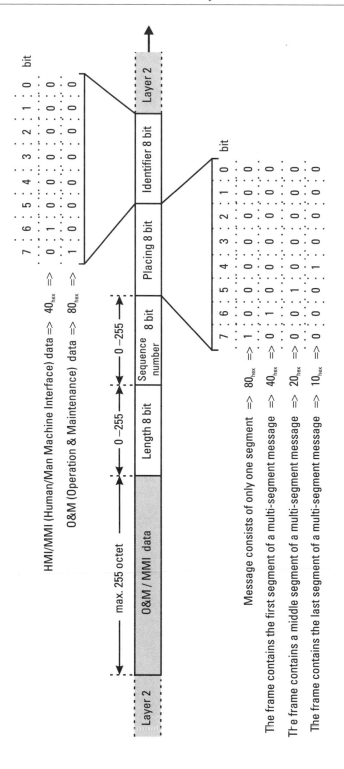

Figure 6.30 Format of Layer 3 on OML (SAPI 62).

distinguishes between HMI data and O&M information. That detail must be noted to ensure the distinction between maintenance communication and data transmission.

The next octet is a placing octet, which indicates to the recipient whether a message is segmented, that is, whether it is made up of a number of submessages. The next parameter is the sequence number, which numbers the segmented messages.

The last octet of the header is a length indicator, which indicates how many octets of O&M data follow.

The O&M communication of the BTS with the BSC is illustrated in Figure 6.31 with the example of a file transfer. All messages shown are I frames. To reduce the clutter of the graphic, the acknowledging RR frames are not shown; normally, they would be present.

HMI Data

The OML also is required to transfer maintenance information to the BTS. The related commands have origins in either the OMC or the BSC.

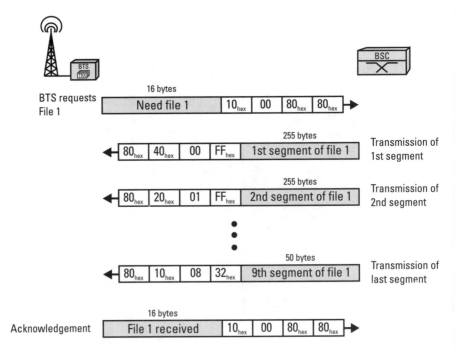

Figure 6.31 File transfer from the BSC to the BTS.

6.4 Bringing an Abis-Interface Into Service

6.4.1 Layer 1

To bring a link between the BTS and the BSC into service, a physical connection first has to be established.

Having a reliable physical link between the BSC and the BTS is the precondition to loading software into the BTS and bringing it up into service. Layer 1 can be tested with any tool capable of measuring bit errors. Another possibility, of course, is to retrieve the link state from the OMC.

6.4.2 Layer 2

When the Layer 1 connection has been established and the hardware on both sides, the BTS and the BSC, is operable, SABME frames can be detected in the time slots for signaling data at both ends, as illustrated in Figure 6.32. Note that these SABMEs are sent by the BSC side. The link establishment procedure then follows.

Data can be transported on the Abis-interface as soon as Layer 2 is operable.

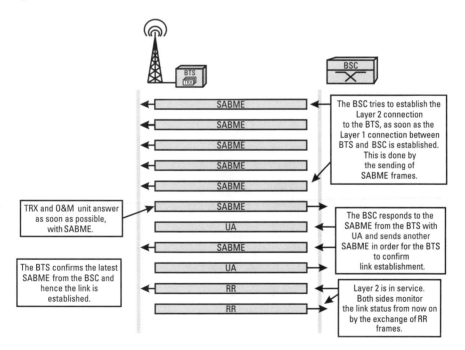

Figure 6.32 An Abis link is brought into service.

7

The Air-Interface of GSM

The Air-interface is the central interface of every mobile system and typically the only one to which a customer is exposed.

The physical characteristics of the Air-interface are particularly important for the quality and success of a new mobile standard. For some mobile systems, only the Air-interface was specified in the beginning, like IS-95, the standard for CDMA. Although different for GSM, the Air-interface still has received special attention. Considering the small niches of available frequency spectrum for new services, the efficiency of frequency usage plays a crucial part. Such efficiency can be expressed as the quotient of transmission rate (kilobits per second) over bandwidth (kilohertz). In other words, how much traffic data can be squeezed into a given frequency spectrum at what cost?

The answer to that question eventually will decide the winner of the recently erupted battle among the various mobile standards.

7.1 The Structure of the Air-Interface in GSM

7.1.1 The FDMA/TDMA Scheme

GSM utilizes a combination of frequency division multiple access (FDMA) and time division multiple access (TDMA) on the Air-interface. That results in a two-dimensional channel structure, which is presented in Figure 7.1. Older standards of mobile systems use only FDMA (an example for such a network is the C-Netz in Germany in the 450 MHz range). In such a pure FDMA system, one specific frequency is allocated for every user during a call. That quickly leads to overload situations in cases of high demand. GSM took into account

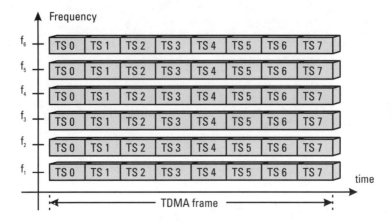

Figure 7.1 The FDMA/TDMA structure of GSM.

the overload problem, which caused most mobile communications systems to fail sooner or later, by defining a two-dimensional access scheme. In fullrate configuration, eight time slots (TSs) are mapped on every frequency; in a halfrate configuration there are 16 TSs per frequency.

In other words, in a TDMA system, each user sends an impulselike signal only periodically, while a user in a FDMA system sends the signal permanently. The difference between the two is illustrated in Figure 7.2. Frequency 1 (f1) in the figure represents a GSM frequency with one active TS, that is, where a signal is sent once per TDMA frame. That allows TDMA to simultaneously serve seven other channels on the same frequency (with fullrate configuration) and manifests the major advantage of TDMA over FDMA (f2).

The spectral implications that result from the emission of impulses are not discussed here. It needs to be mentioned that two TSs are required to support duplex service, that is, to allow for simultaneous transmission and reception. Considering that Figures 7.1 and 7.2 describe the downlink, one can imagine the uplink as a similar picture on another frequency.

GSM uses the modulation technique of Gaussian minimum shift keying (GMSK). GMSK comes with a narrow frequency spectrum and theoretically no amplitude modulation (AM) part. The Glossary provides more details on GMSK.

7.1.2 Frame Hierarchy and Frame Numbers

In GSM, every impulse on frequency 1, as shown in Figure 7.2, is called a burst. Therefore, every burst shown in Figure 7.2 corresponds to a TS. Eight bursts or TSs, numbered from 0 through 7, form a TDMA frame.

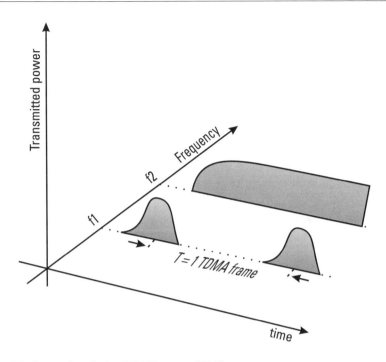

Figure 7.2 Spectral analysis of TDMA versus FDMA.

In a GSM system, every TDMA frame is assigned a fixed number, which repeats itself in a time period of 3 hours, 28 minutes, 53 seconds, and 760 milliseconds. This time period is referred to as hyperframe. Multiframe and superframe are layers of hierarchy that lie between the basic TDMA frame and the hyperframe. Figure 7.3 presents the various frame types, their periods, and other details, down to the level of a single burst as the smallest unit.

Two variants of multiframes, with different lengths, need to be distinguished. There is the 26-multiframe, which contains 26 TDMA frames with a duration of 120 ms and which carries only traffic channels and the associated control channels. The other variant is the 51-multiframe, which contains 51 TDMA frames with a duration of 235.8 ms and which carries signaling data exclusively. Each superframe consists of twenty-six 51-multiframes or fifty-one 26-multiframes. This definition is purely arbitrary and does not reflect any physical constraint. The frame hierarchy is used for synchronization between BTS and MS, channel mapping, and ciphering.

Every BTS permanently broadcasts the current frame number over the synchronization channel (SCH) and thereby forms an internal clock of the BTS. There is no coordination between BTSs; all have an independent clock, except for synchronized BTSs (see *synchronized handover* in the Glossary). An

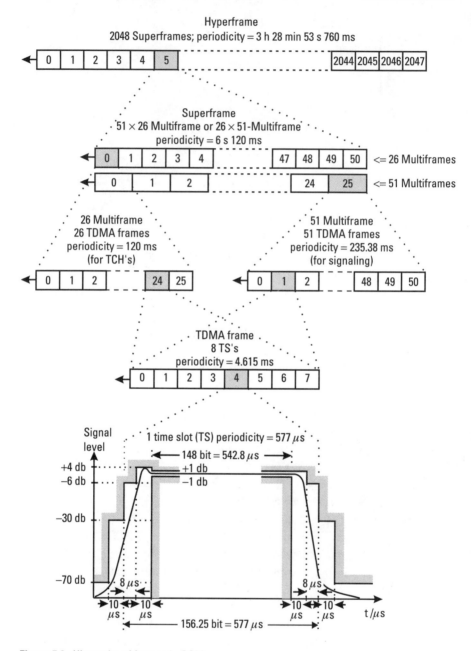

Figure 7.3 Hierarchy of frames in GSM.

MS can communicate with a BTS only after the MS has read the SCH data, which informs the MS about the frame number, which in turn indicates the chronologic sequence of the various control channels. That information is very important, particularly during the initial access to a BTS or during handover.

Consider this example: an MS sends a channel request to the BTS at a specific moment in time, let's say frame number Y ($t = FN\ Y$). The channel request is answered with a channel assignment, after being processed by the BTS and the BSC. The MS finds its own channel assignment among all the other ones, because the channel assignment refers back to frame number Y.

The MS and the BTS also need the frame number information for the ciphering process. The hyperframe with its long duration was only defined to support ciphering, since by means of the hyperframe, a frame number is repeated only about every three hours. That makes it more difficult for hackers to intercept a call.

7.1.3 Synchronization Between Uplink and Downlink

For technical reasons, it is necessary that the MS and the BTS do not transmit simultaneously. Therefore, the MS is transmitting three timeslots after the BTS. The time between sending and receiving data is used by the MS to perform various measurements on the signal quality of the receivable neighbor cells.

As shown in Figure 7.4, the MS actually does not send exactly three timeslots after receiving data from the BTS. Depending on the distance between the two, a considerable propagation delay needs to be taken into account. That propagation delay, known as timing advance (TA), requires the MS to transmit its data a little earlier as determined by the "three timeslots delay rule."

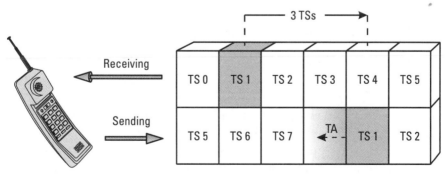

The actual point in time of the transmission is shifted by the Timing Advance

Figure 7.4 Receiving and sending from the perspective of the MS.

The larger the distance between the MS and the BTS is, the larger the TA is. More details are provided in the Glossary under *TA*.

7.2 Physical Versus Logical Channels

Because this text frequently uses the terms *physical channel* and *logical channel*, the reader should be aware of the differences between them.

- Physical channels are all the available TSs of a BTS, whereas every TS corresponds to a physical channel. Two types of channels need to be distinguished, the halfrate channel and the fullrate channel. For example, a BTS with 6 carriers, as shown in Figure 7.1, has 48 (8 times 6) physical channels (in fullrate configuration).
- Logical channels are piggybacked on the physical channels. Logical channels are, so to speak, laid over the grid of physical channels. Each logical channel performs a specific task.

Another aspect is important for the understanding of logical channels: during a call, the MS sends its signal periodically, always in a TDMA frame at the same burst position and on the same TS to the BTS (e.g., always in TS number 3). The same applies for the BTS in the reverse direction.

It is important to understand the mapping of logical channels onto available TSs (physical TSs)—which will be discussed later—because the channel mapping always applies to the same TS number of consecutive TDMA frames. (The figures do not show the other seven TSs.)

7.3 Logical-Channel Configuration

Firstly, the distinction should be made between traffic channels (TCHs) and control channels (CCHs). Distinguishing among the different TCHs is rather simple, since it only involves the various bearer services. Distinguishing among the various CCHs necessary to meet the numerous signaling needs in different situations, however, is more complex. Table 7.1 summarizes the CCH types, and the Glossary provides a detailed description of each channel and its tasks. Note that, with three exceptions, the channels are defined for either downlink or uplink only.

Table 7.1
Signaling Channels of the Air-Interface

Name	Abbreviation	Task
Frequency correction channel (DL)	FCCH	The "lighthouse" of a BTS
Synchronization channel (DL)	SCH	PLMN/base station identifier of a BTS plus synchronization information (frame number)
Broadcast common control channel (DL)	BCCH	To transmit system information 1–4, 7-8 (differs in GSM, DCS1800, and PCS1900)
Access grant channel (DL)	AGCH	SDCCH channel assignment (the AGCH carries IMM_ASS_CMD)
Paging channel (DL)	PCH	Carries the PAG_REQ message
Cell broadcast channel (DL)	CBCH	Transmits cell broadcast messages (see Glossary entry *CB*)
Standalone dedicated control channel	SDCCH	Exchange of signaling information between MS and BTS when no TCH is active
Slow associated control channel	SACCH	Transmission of signaling data during a connection (one SACCH TS every 120 ms)
Fast associated control channel	FACCH	Transmission of signaling data during a connection (used only if necessary)
Random access channel (UL)	RACH	Communication request from MS to BTS

Note: DL = downlink direction only; UL = uplink direction only.

7.3.1 Mapping of Logical Channels Onto Physical Channels

In particular, the downlink direction of TS 0 of the BCCH-TRX is used by various channels. The following channel structure can be found on TS 0 of a BCCH-TRX, depending on the actual configuration:

- FCCH;
- SCH;
- BCCH information 1–4;
- Four SDCCH subchannels (optional);
- CBCH (optional).

This multiple use is possible because the logical channels can time-share TS 0 by using different TDMA frames. A remarkable consequence of the approach is that, for example, the FCCH or the SCH of a BTS is not broadcast permanently but is there only from time to time. Time sharing of the same TS is not limited to FCCH and SCH but is widely used. Such an approach naturally results in a lower transmission capacity, which is still sufficient to convey all necessary signaling data. Furthermore, it is possible to combine up to four physical channels in consecutive TDMA frames to a block, so that it is possible for the same SDCCH to use the same physical channel in four consecutive TDMA frames, as illustrated in Figure 7.5. On the other hand, an SDCCH subchannel has to wait for a complete 51-multiframe before it can be used again.

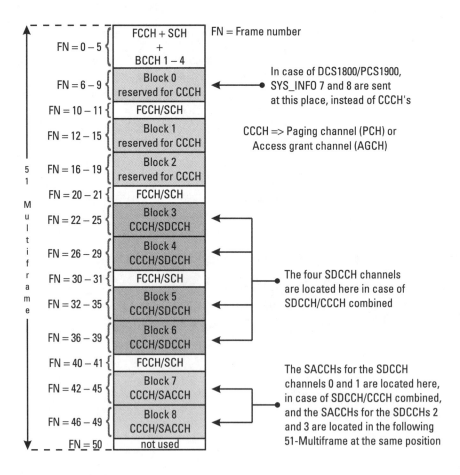

Figure 7.5 Example of the mapping of logical channels.

That clarifies another reason for the frame hierarchy of GSM. The structure of the 51-multiframe defines at which moment in time a particular control channel (logical channel) can use a physical channel (it applies similarly to the 26-multiframe).

Detailed examples are provided in Figure 7.6, for the downlink, and in Figure 7.7, for the uplink. The figures show a possible channel configuration for all eight TSs of a TRX. Both show a 51-multiframe in TSs 0 and 1, with a cycle time of 235.8 ms. Each of the remaining TSs, 2 through 7, carries two 26-multiframes, with a cycle time of 2 · 120 ms = 240 ms. That explains the difference in length between TS 0 and TS 1 on one hand and TS 2 through TS 7 on the other.

Figures 7.6 and 7.7 show that a GSM 900 system can send the BCCH SYS-INFO 1–4 only once per 51-multiframe. That BCCH information tells the registered MSs all the necessary details about the channel configuration of a BTS. That includes at which frame number a PAG_REQ is sent on the PCH and which frame numbers are available for the RACH in the uplink direction. The Glossary provides more details on the content of BCCH SYS-INFO 1–4.

The configuration presented in Figures 7.6 and 7.7 contains 11 SDCCH subchannels: 3 on TS 0 and another 8 on TS 1. SDCCH 0, 1, ... refers to the SDCCH subchannel 0, 1, ... on TS 0 or TS 1. The channel configuration presented in the figures also contains a CBCH on TS 0. Note that the CBCH will always be exactly at this position of TS 0 or TS 1 and occupies the frame numbers 8–11. The CBCH reduces, in both cases, the number of available SDCCH subchannels (that is why SDCCH/2 is missing in the example).

The configuration, as presented here, is best suited for a situation in which a high signaling load is expected while only a relatively small amount of payload is executed. Only the TSs 2 through 7 are configured for regular full-rate traffic.

The shaded areas indicate the so-called idle frame numbers, that is, where no information transfer occurs.

7.3.2 Possible Combinations

The freedom to define a channel configuration is restricted by a number of constraints. When configuring a cell, a network operator has to consider the peculiarities of a service area and the frequency situation, to optimize the configuration. Experience with the average and maximum loads that are expected for a BTS and how the load is shared between signaling and payload is an important factor for such consideration.

GSM 05.02 provides the following guidelines, which need to be taken into account when setting up control channels.

	FN	TS 0	TS 1	FN	TS 2	TS 3 - 6	TS 7
	0	FCCH	SDCCH 0	0	TCH		TCH
	1	SCH	SDCCH 0	1	TCH		TCH
	2	BCCH 1	SDCCH 0	2	TCH		TCH
	3	BCCH 2	SDCCH 0	3	TCH		TCH
	4	BCCH 3	SDCCH 1	4	TCH		TCH
	5	BCCH 4	SDCCH 1	5	TCH		TCH
	6	AGCH/PCH	SDCCH 1	6	TCH		TCH
	7	AGCH/PCH	SDCCH 1	7	TCH	2	TCH
	8	AGCH/PCH	SDCCH 2	8	TCH	6	TCH
	9	AGCH/PCH	SDCCH 2	9	TCH		TCH
	10	FCCH	SDCCH 2	10	TCH	M	TCH
	11	SCH	SDCCH 2	11	TCH	u	TCH
	12	AGCH/PCH	SDCCH 3	12	SACCH	l	SACCH
	13	AGCH/PCH	SDCCH 3	13	TCH	t	TCH
	14	AGCH/PCH	SDCCH 3	14	TCH	i	TCH
	15	AGCH/PCH	SDCCH 3	15	TCH	f	TCH
	16	AGCH/PCH	SDCCH 4	16	TCH	r	TCH
	17	AGCH/PCH	SDCCH 4	17	TCH	a	TCH
5	18	AGCH/PCH	SDCCH 4	18	TCH	m	TCH
1	19	AGCH/PCH	SDCCH 4	19	TCH	e	TCH
	20	FCCH	SDCCH 5	20	TCH		TCH
M	21	SCH	SDCCH 5	21	TCH		TCH
u	22	SDCCH 0	SDCCH 5	22	TCH		TCH
l	23	SDCCH 0	SDCCH 5	23	TCH		TCH
t	24	SDCCH 0	SDCCH 6	24	TCH		TCH
i	25	SDCCH 0	SDCCH 6	25			
f	26	SDCCH 1	SDCCH 6	0	TCH		TCH
r	27	SDCCH 1	SDCCH 6	1	TCH		TCH
a	28	SDCCH 1	SDCCH 7	2	TCH		TCH
m	29	SDCCH 1	SDCCH 7	3	TCH		TCH
e	30	FCCH	SDCCH 7	4	TCH		TCH
	31	SCH	SDCCH 7	5	TCH		TCH
	32	CBCH	SACCH 0	6	TCH		TCH
	33	CBCH	SACCH 0	7	TCH	2	TCH
	34	CBCH	SACCH 0	8	TCH	6	TCH
	35	CBCH	SACCH 0	9	TCH		TCH
	36	SDCCH 3	SACCH 1	10	TCH	M	TCH
	37	SDCCH 3	SACCH 1	11	TCH	u	TCH
	38	SDCCH 3	SACCH 1	12	SACCH	l	SACCH
	39	SDCCH 3	SACCH 1	13	TCH	t	TCH
	40	FCCH	SACCH 2	14	TCH	i	TCH
	41	SCH	SACCH 2	15	TCH	f	TCH
	42	SACCH 0	SACCH 2	16	TCH	r	TCH
	43	SACCH 0	SACCH 2	17	TCH	a	TCH
	44	SACCH 0	SACCH 3	18	TCH	m	TCH
	45	SACCH 0	SACCH 3	19	TCH	e	TCH
	46	SACCH 1	SACCH 3	20	TCH		TCH
	47	SACCH 1	SACCH 3	21	TCH		TCH
	48	SACCH 1		22	TCH		TCH
	49	SACCH 1		23	TCH		TCH
	50			24	TCH		TCH
				25			

Figure 7.6 Example of the downlink part of a fullrate channel configuration of FCCH/SCH + CCCH + SDCCH/4 + CBCH on TS 0, SDCCH/8 on TS 1, and TCHs on TSs 2–7. The missing SACCHs on TS 0 and TS 1 can be found in the next multiframe, which is not shown here. There is no SDCCH/2 on TS 0, because of the CBCH.

The Air-Interface of GSM

FN	TS 0	TS 1	FN	TS 2	TS 3 - 6	TS 7
0	SDCCH 3	SACCH 1	0	TCH		TCH
1	SDCCH 3	SACCH 1	1	TCH		TCH
2	SDCCH 3	SACCH 1	2	TCH		TCH
3	SDCCH 3	SACCH 1	3	TCH		TCH
4	RACH	SACCH 2	4	TCH		TCH
5	RACH	SACCH 2	5	TCH		TCH
6	SACCH 2	SACCH 2	6	TCH		TCH
7	SACCH 2	SACCH 2	7	TCH	2	TCH
8	SACCH 2	SDCCH 3	8	TCH	6	TCH
9	SACCH 2	SACCH 3	9	TCH		TCH
10	SACCH 3	SACCH 3	10	TCH	M	TCH
11	SACCH 3	SACCH 3	11	TCH	u	TCH
12	SACCH 3		12	SACCH	l	SACCH
13	SACCH 3		13	TCH	t	TCH
14	RACH		14	TCH	i	TCH
15	RACH	SDCCH 0	15	TCH	f	TCH
16	RACH	SDCCH 0	16	TCH	r	TCH
17	RACH	SDCCH 0	17	TCH	a	TCH
18	RACH	SDCCH 0	18	TCH	m	TCH
19	RACH	SDCCH 1	19	TCH	e	TCH
20	RACH	SDCCH 1	20	TCH		TCH
21	RACH	SDCCH 1	21	TCH		TCH
22	RACH	SDCCH 1	22	TCH		TCH
23	RACH	SDCCH 2	23	TCH		TCH
24	RACH	SDCCH 2	24	TCH		TCH
25	RACH	SDCCH 2	25			
26	RACH	SDCCH 2	0	TCH		TCH
27	RACH	SDCCH 3	1	TCH		TCH
28	RACH	SDCCH 3	2	TCH		TCH
29	RACH	SDCCH 3	3	TCH		TCH
30	RACH	SDCCH 3	4	TCH		TCH
31	RACH	SDCCH 4	5	TCH		TCH
32	RACH	SDCCH 4	6	TCH		TCH
33	RACH	SDCCH 4	7	TCH	2	TCH
34	RACH	SDCCH 4	8	TCH	6	TCH
35	RACH	SDCCH 5	9	TCH		TCH
36	RACH	SDCCH 5	10	TCH	M	TCH
37	SDCCH 0	SDCCH 5	11	TCH	u	TCH
38	SDCCH 0	SDCCH 5	12	SACCH	l	SACCH
39	SDCCH 0	SDCCH 6	13	TCH	t	TCH
40	SDCCH 0	SDCCH 6	14	TCH	i	TCH
41	SDCCH 1	SDCCH 6	15	TCH	f	TCH
42	SDCCH 1	SDCCH 6	16	TCH	r	TCH
43	SDCCH 1	SDCCH 7	17	TCH	a	TCH
44	SDCCH 1	SDCCH 7	18	TCH	m	TCH
45	RACH	SDCCH 7	19	TCH	e	TCH
46	RACH	SDCCH 7	20	TCH		TCH
47		SACCH 0	21	TCH		TCH
48		SACCH 0	22	TCH		TCH
49		SACCH 0	23	TCH		TCH
50		SACCH 0	24	TCH		TCH
			25			

Figure 7.7 Example of the uplink part of a fullrate channel configuration. RACHs can be found only on TS 0 of the designated frame numbers. The missing SACCHs on TS 0 and TS 1 can be found in the next multiframe, which is not shown here.

- The FCCH and the SCH are always sent in TS 0 of the BCCH carrier at specific frame numbers (see Figure 7.5).
- The BCCH, RACH, PCH, and AGCH also must be assigned only to the BCCH carrier. These channels, however, allow for assignment to all even-numbered TSs, e.g., 0, 2, 4, and 6, as well as to various frame numbers.

In practice, two configurations are mainly used, which can be combined if necessary (compare Figure 7.6 and Figure 7.7):

- FCCH + SCH + BCCH + CCCH // SDCCH/8 addresses a channel configuration in which no SDCCH subchannels are available on TS 0. Eight such SDCCH subchannels are defined on TS 1. In that case, TS 1 obviously is not available as a traffic channel.
- FCCH + SCH + BCCH + CCCH + SDCCH/4 addresses a channel configuration in which all control channels are assigned to TS 0, in particular, to have TS 1 available to carry payload traffic. Because TS 0 needs to be used by the other control channels, too, it is possible to establish only four SDCCH subchannels, that is, only half the number compared to the preceding configuration.

A channel configuration is always related to a single TS and not to a complete TRX. It is not possible to combine traffic channels and SDCCHs. If necessary, a TS can be "sacrificed" to allow for additional SDCCHs.

7.4 Interleaving

The preceding descriptions were made under an assumption that is not valid for the Air-interface of GSM. That assumption is that data are transmitted in the order they were generated or received, that is, the first bit of the first (spoken) word is sent first. That is not the case for the Air-interface of GSM. Figure 7.8 illustrates the process of interleaving smaller packages of 456 bits over a larger time period, that is, distributing them in separate TSs. How the packets are spread depends on the type of application the bits represent. Signaling traffic and packets of data traffic are spread more than voice traffic. The whole process is referred to as interleaving.

The goal of interleaving is to minimize the impact of the peculiarities of the Air-interface that account for rapid, short-term changes of the quality of the

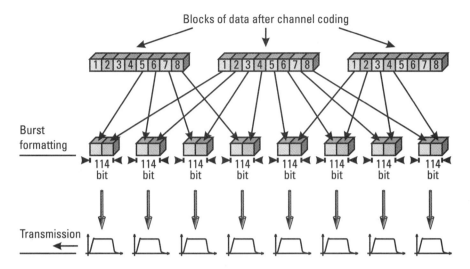

Figure 7.8 Interleaving of speech traffic.

transmission channel. It is possible that a particular channel is corrupted for a very short period of time and all the data sent during that time are lost. That could lead to loss of complete data packets of n times 114 bits. Interleaving does not prevent loss of bits, and if there is a loss, the same number of bits are lost. However, because of interleaving, the lost bits are part of several different packets, and each packet loses only a few bits out of a larger number of bits. The idea is that those few bits can be recovered by error-correction mechanisms.

7.5 Signaling on the Air-Interface

7.5.1 Layer 2 LAPD$_m$ Signaling

The only GSM-specific signaling of OSI Layers 1 and 2 can be found on the Air-interface, where LAPD$_m$ signaling is used. The other interfaces of GSM use already defined protocols, like LAPD and SS7.

The abbreviation LAPD$_m$ suggests that it refers to a protocol closely related to LAPD, which is correct. The "m" stands for "modified" and the frame structure already shows the closeness to LAPD. The modified version of LAPD is an optimized version for the GSM Air-interface and was particularly tailored to deal with the limited resources and the peculiarities of the radio link. All dispensable parts of the LAPD frame were removed to save resources. The

LAPD$_m$ frame, in particular, lacks the TEI, the FCS, and the flags at both ends. The LAPD$_m$ frame does not need those parts, since their task is performed by other GSM processes. The task of the FCS, for instance, to a large extent, is performed by channel coding/decoding.

7.5.1.1 The Three Formats of the LAPD$_m$ Frame

Figure 7.9 is an overview of the frame structure of LAPD$_m$. Three different formats of identical length (23 bytes) are defined; their respective uses depend on the type of information to be transferred.

- A-format. A frame in the A-format generally can be sent on any DCCH in both directions, uplink and downlink. The A-format frame is sent as a fill frame when no payload is available on an active connection, for example, in the short time period immediately after the traffic channel is connected.

- B-format. The B-format is used on the Air-interface to transport the actual signaling data; hence, every DCCH and every ACCH use this format. The maximum length of the Layer 3 information to be carried is restricted, depending on the channel type (SDCCH, FACCH, SACCH). This value is defined per channel type by the constant N201. If the information to be transmitted requires less space, this space has to be filled with fill-in octets.

- Bbis-format. For transmission of BCCH, PCH, and AGCH. There is no header in the Bbis-format that would allow for addressing or frame identification. Addressing is not necessary, since BCCH, PCH, and AGCH are CCCHs, in which addressing is not required. In contrast to the DCCH, the CCCH transports only point-to-multipoint messages.

Both frame types, the A-format and the B-format, are used in both directions, uplink and downlink. The Bbis format is required for the downlink only.

Also noteworthy is the relationship for signaling information between the maximum frame length of an LAPD$_m$ frame (= 23 byte $\equiv$ 184 bit) and the number of input bits for channel coding (= 184 bit).

7.5.1.2 The Header of an LAPD$_m$ Frame

The Address Field

The address field starts with the bits EA and C/R, which perform the same tasks as the parameters with the same names in an LAPD frame. The same applies for SAPI, which takes on different values over the Air-interface than on

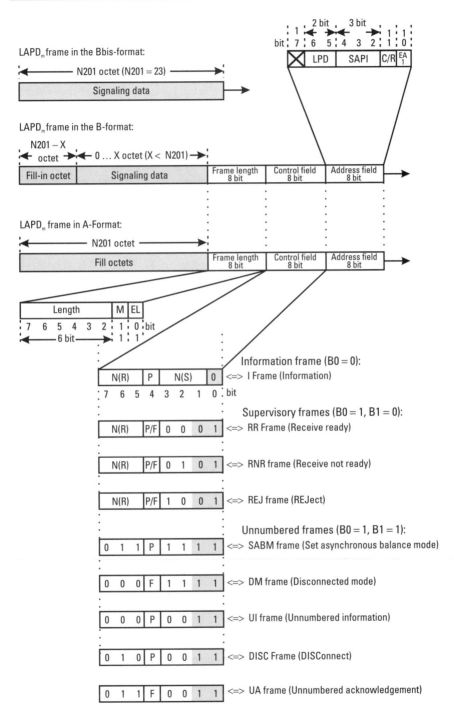

Figure 7.9 Frame format and frame type of $LAPD_m$.

the Abis-interface. Table 7.2 lists the possible values for SAPIs on the Air-interface and their uses. SAPI = 0 is used for all messages that deal with CC, MM, and RR, while SAPI = 3 is used for messages related to supplementary services and the SMS.

Furthermore, the address field of an $LAPD_m$ frame contains the 2-bit-long link protocol discriminator (LPD), which in GSM is, with one exception, always coded with 00_{bin}. The exception is the cell broadcast service (CBS), where LPD = 01_{bin}.

Control Field

The control field of an $LAPD_m$ frame is identical to that of an LAPD frame modulo 8. It defines the frame type and contains, in the case of I frames, the counters for N(S) and N(R); in the case of supervisory frames, it contains only N(R).

The *frame length indicator field* consists of three parts:

- Bit 0, the EL-bit. The EL-bit indicates if the current octet is the last one of the frame length indicator field. When this bit is set to 1, then another length indication octet follows, if set to 0, this octet is the last one. GSM does not allow the frame length indicator field to exceed one octet, and hence, the value of the EL-bit is always zero. GSM may change this restriction, if future applications require a different length.

- Bit 1, the M-bit. If entire messages are longer than the data field of the $LAPD_m$ frames allows, the information has to be partitioned and transmitted in consecutive frames. The M-bit is used in such a situation to indicate that the message was segmented and that further frames belonging to the same messages have to be expected. The M-bit of the last segment is set to zero, as illustrated in Figure 7.10.

- Bits 2–7, the length indicator. This field indicates the actual length of the information field. The value range is from zero to N201.

Table 7.2
Possible Values of SAPI on the Air-Interface

SAPI (Decimal)	Meaning
0	RR, MM, CC
3	SMS, SS

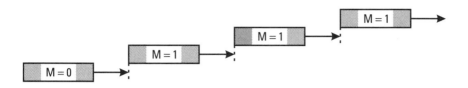

Figure 7.10 Segmentation in LAPD$_m$.

Information Field

For all three frame formats, the information field that carries signaling data consists of N201 octets, where N201 represents a value that is different for the various channel types (see *N201* in the Glossary). How many of the octets—in the case of a B-format—are actually part of Layer 3 depends on the data to be transported. It is important to note that all unused octets in case of the B-format and all octets of the A-format are so-called fill-in octets, which are coded in a precisely defined pattern. This bit pattern is different for uplink and downlink. If, for example, an SDCCH frame contains only 18 bytes of data, the remaining two bytes are occupied with fill-in octets (note that N201 for the SDCCH has a value of 20).

7.5.1.3 Differences Between LAPD and LAPD$_m$

The differences between LAPD and LAPD$_m$ are as follows:

- LAPD$_m$ frames exist in modulo 8 format only. Their control field, therefore, is always 1 octet long. The N(S) and the N(R) are in the range 0 to 7. That theoretically restricts the maximum number of unacknowledged I frames to seven.
- The address field of LAPD$_m$ is only 1 octet long and does not contain a TEI. The reason is that when a channel is already assigned, the connection on the Air-interface is always a point-to-point connection. Several simultaneous users, for example, on a terrestrial point-to-multipoint connection, do not exist, which makes the TEI superfluous.
- LAPD$_m$ frames do not contain an FCS, because channel coding and interleaving of Layer 1 already provide data security.
- LAPD$_m$ frames do not have a flag to indicate the start and end of a frame. That functionality is provided on the Air-interface by Layer 1, in particular by the burst segmentation.
- Unlike in LAPD, SABM frames and UA frames of LAPD$_m$ may even carry Layer 3 data. That saves time during connection setup.

- The maximum lengths of LAPD and LAPD$_m$ frames are very different. While LAPD frames can transport up to 260 octets of signaling data, LAPD$_m$ allows for only 23 octets. If a larger amount of data needs to be transported, segmentation has to be applied.
- LAPD$_m$ frames do not contain a length indicator (Layer 2).
- In LAPD, no fill-in octets are used when the data area is not completely occupied with signaling data.

7.5.1.4 Frame Types of LAPD$_m$

Fewer frame types are defined for the LAPD$_m$ protocol than for LAPD. The XID frame and the FRMR frame are missing in LAPD$_m$. Both frames are used for specific tasks and are not necessary in LAPD$_m$. Table 7.3 lists the frame types of LAPD$_m$ and their specific uses. As for LAPD, it is distinguished whether a frame is used to carry a command, a response, or both. LAPD$_m$ follows the definition of LAPD, that is, the P/F bit and the C/R bits are used the same way for both protocols.

Table 7.3
Frame Types of the Air-Interface

Name	Command Frame?	Answer Frame?	Possible Values of Control Field (Hex)
I-frame group:			
I	Yes	No	(0X), (2X), (4X), (6X), (8X) if even, then I frame
Supervisory-frame group			
RR	Yes	Yes	(1X)
RNR	Yes	Yes	(5X)
REJ	Yes	Yes	(9X)
Unnumbered-frame group			
DISC	Yes	No	(53) because P bit is always 1
UI	Yes	No	(03) because P bit always 0
DM	No	Yes	(0F), (1F)
SABME	Yes	No	(7F) because P bit always 1
UA	No	Yes	(73) because F bit always 1

7.5.2 Layer 3

Figure 7.11 illustrates the Layer 3 format on the Air-interface.

7.5.2.1 Protocol Discriminator

The 4-bit-long protocol discriminator (PD) is used on the Air-interface to classify all messages into groups and allows, within Layer 3, the addressing of various users, just as the message discriminator does on the Abis-interface. Every message is nonambiguously assigned to a PD or service class. A distinction between transparent and nontransparent services is possible at the same time. Supplementary services and the SMS are special, because they do not belong to CC but are still sent with the same PD. Table 7.4 lists all PDs and their service classes.

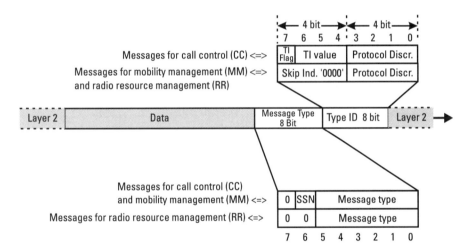

Figure 7.11 The Layer 3 format on the Air-interface.

Table 7.4
Protocol Discriminators on the Air-Interface

PD	Service Class
06	RR (radio resource management)
05	MM (mobility management)
03	CC (call control)
	SS (supplementary services)
	SMS (short-message services)

7.5.2.2 Radio Resource Management

Messages in the area of RR are necessary to manage the logical as well as the physical channels on the Air-interface. Depending on the message type, processing of RR messages is performed by the MS, in the BSS, or even in the MSC. Involvement of the BSS distinguishes RR from MM and CC.

7.5.2.3 Mobility Management

MM uses the channels that RR provides, to transparently exchange data between the MS and the NSS. From a hierarchical perspective, the MM lies above the RR, because MM data already are user data. The BSS does not, with a few exceptions, process MM messages. A typical application of MM is location update.

7.5.2.4 Call Control

Like MM, CC uses the connection that RR provides for information exchange. In contrast to MM, which is used only to maintain the mobility of a subscriber, CC is a real application that at the same time provides an interface to ISDN. (The relation between CC and ISDN is discussed in Chapter 10.)

7.5.2.5 Transaction Identifier and Skip Indicator

In CC, the PD is followed by the transaction identifier (TI); in MM and RR, the PD is followed by the skip indicator. The skip indicator in RR and MM messages is a 4-bit-long, fixed coded dummy value with 0000_{bin}. No specific task is assigned to the skip indicator. Messages in which the skip indicator is not 0000_{bin} are ignored by the receiver and indicate a transmission error.

The 4-bit-long TI, on the other hand, can distinguish among several simultaneous transactions of one MS. The format of the TI, shown in Figure 7.11, is separated into the TI flag and the TI value.

The TI flag (bit 7) is used to distinguish between the initiating side and the responding side of a transaction. For the initiating side, the TI flag is set to 0; for the responding side, it has a value of 1. Hence, in a MOC, the TI flags of all CC messages sent from the MS are set to 0. Correspondingly, the TI flags of all CC messages sent from the NSS have a value of 1. In a MTC, the reciprocal applies.

The initiating side also assigns the TI value, which can be in the range of 0 through 6. One TI value is assigned for every transaction, where it is allowed that the MS and the NSS assign the same TI value to different transactions. The TI flag is used in that case to avoid ambiguity. Several simultaneous transactions are allowed only in the CC protocol, so neither MM nor RR require a TI.

Figure 7.12 illustrates this relation. When the MS is involved in an active call, it places the call on hold and sets up the second call.

7.5.2.6 The Message Type

The value of the protocol discriminator also determines the format of this octet (see Figure 7.11). The first six bits (bits 0 to 5) indicate the message type itself. Section 7.5.2.7 explains all the message types of the Air-interface in more detail. The format of its parameters is shown in Figure 7.13. A distinction is made between mandatory and optional parameters with fixed or variable length, which requires an information element identifier and/or a length indicator.

A special task takes bit number six of the message type. While bits 6 and 7 of the RR are fix-coded with 00_{bin}, bit 6 of MM and CC is held by the send sequence number. No special task is assigned to the send sequence number of MM and CC messages in the downlink direction and is, hence, fix-coded with 0. In the uplink direction, however, the send sequence number of MM and CC messages toggles between a value of 0 and 1. Figure 7.14 provides an example. Note that the send sequence number toggles simultaneously for both CC and MM. The change of the value of the send sequence number is significant for

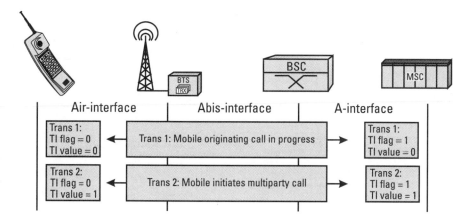

Figure 7.12 Task of the TI in case of several simultaneous CC transactions.

protocol testing, because of two possible values in the uplink direction of MM and CC messages.

7.5.2.7 The Message Type, Bits 0 Through 5

Tables 7.5, 7.6, 7.7, and 7.8 list all the messages that are defined on the Air-interface, together with brief descriptions of their tasks. The messages are ordered according to protocol groups into RR, MM, CC, and supplementary services. Note that two different hexadecimal values for the message type are possible, because of the send sequence number in bit 6 of the message type of MM and CC messages.

The characters in uppercase indicate the abbreviations used in the description.

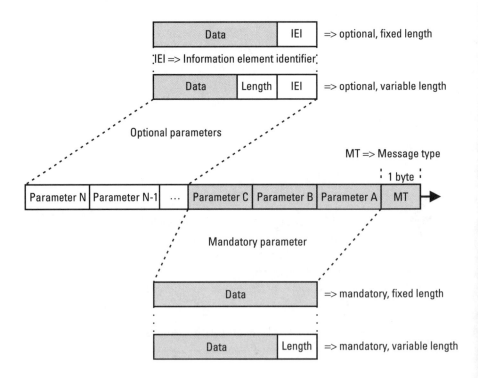

Figure 7.13 Parameter format and Air-interface signaling.

The Air-Interface of GSM 111

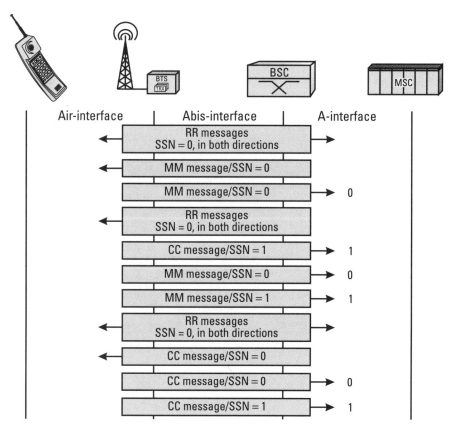

Figure 7.14 Use of the send sequence number.

Table 7.5
Radio Resource Management (Skip Indicator/Protocol Discriminator = 06)

ID (Hex)	Name	Direction	Description
-/-	CHANnel REQuest	MS → BTS	CHAN_REQ is a request of an MS for a channel when in the idle state. Although only 1 byte long this message already contains the reason for the connection request (answer to PAGING, Emergency Call, etc.) and an identifier for the channel type that the MS prefers. The CHAN_REQ has no hexadecimal message type, because the message does not conform to the regular format and is sent via an access burst.

Table 7.5 (continued)

ID (Hex)	Name	Direction	Description
-/-	HaNDover ACCess	MS → BTS	The MS sends consecutive HND_ACC messages on a new traffic channel for every handover (synchronized and nonsynchronized). The only exception is the intra-BTS handover via ASS_CMD. Like the CHAN_REQ, the HND_ACC does not follow the standard format and is sent in an access burst to the BTS. The handover reference is the only information that HND_ACC contains and is assigned with the HND_CMD message to allow for identification of the "correct" MS during BTS access.
02	SYStem INFOrmation 2bis	BTS → MS	The data area of the SYS_INFO 2 is not large enough to allow for distinction of the larger number of channels of DCS 1800, PCS 1900, and also GSM900 with extended band. Hence, SYS_INFO 2bis and 2ter were defined to broadcast, in particular, the frequencies of the neighbor cells, which do not fit into SYS_INFO 2
03	SYStem INFOrmation 2ter	BTS → MS	See SYS_INFO 2bis
05	SYStem INFOrmation 5bis	BTS → MS	The same restrictions for SYS_INFO 2 also apply to SYS_INFO 5, which had to be extended by SYS_INFO 5bis and 5ter to accommodate the greater number of channels of DCS 1800, PCS 1900, and GSM900 with extended band. Hence, SYS_INFO 5bis and 5ter mainly transport the BCCH frequencies of the neighboring cells, which do not fit into SYS_INFO 5. The messages are sent to the MS over the SACCH when an active connection exists.
06	SYStem INFOrmation 5ter	BTS → MS	See SYS_INFO 5bis
0A	PARTial RELease	BTS → MS	When an MS has activated two radio channels at the same time, and CC wants to release one channel, a PART_REL message is sent. For the time being, this is defined only for two halfrate channels.
0D	CHANnel RELease	BTS → MS	The CHAN_REL message is used when a connection is disconnected, to release the radio resources on the air interface. Cause 0 is used for normal clearing; for abnormal clearing, for instance, cause 1 is used.
0F	PARTial RELease COMplete	MS → BTS	With this message, the MS confirms receipt and processing of a PART_REL message.

Table 7.5 (continued)

ID (Hex)	Name	Direction	Description
10	CHANnel MODe MODify	BTS → MS	CHAN_MOD_MOD is sent by the network to the MS, to modify the transmission parameters of Layer 1 (change the transmission rate).
12	RR STATUS	MS ↔ BTS	A RR_STATUS message with an appropriate error cause is sent when one side receives an RR that has an error in Layer 3. These kind of protocol errors happen, for example, in case of bit errors on the air interface.
13	CLASSmark ENQuiry	BTS → MS	The network requests the technical identification (power class, available encryption algorithms A5/X, SMS capability, etc.) from the MS. The network expects a CLASS_CHANGE message as a response.
14	FREQuency REDEFinition	BTS → MS	The FREQ_REDEF message allows the network to change the configuration of an existing connection, e.g., the hopping sequence in frequency hopping.
15	MEASurement REPort	MS → BTS	MEAS_REP transfers the current measurement results of the MS to the BTS (uplink measurements). These measurements contain the sending levels of the serving cell and of the neighboring cells. In the case of an active connection, a MEAS_REP is sent to the BTS every 480 ms via the SACCH. The BTS forwards the MEAS_REP to the BSC, embedded in its own measurement results (MEAS_RES).
16	CLASSmark CHANGE	MS → BTS	The MS sends this message when the classmark changes (e.g., when a handheld phone is connected to a booster in a car) or when a request is made by the network (CLASS_ENQ). It contains the current technical capabilities of the MS.
17	CHANnel MODe MODify ACKnowledge	MS → BTS	The MS confirms with CHAN_MOD_MOD_ACK the change to another transmission mode that was requested with CHAN_MOD_MOD.
18	SYStem INFOrmation 8	BTS → MS	See SYS_INFO 7.
19	SYStem INFOrmation 1	BTS → MS	Contains the access rights and frequencies of a BTS. The Glossary provides an example for a BCCH/SYS_INFO 1.
1A	SYStem INFOrmation 2	BTS → MS	Transmission of neighbor cell frequencies, access rights (e.g., access control class), and network color code (NCC). The Glossary provides an example of BCCH/SYS_INFO 2.

Table 7.5 (continued)

ID (Hex)	Name	Direction	Description
1B	SYStem INFOrmation 3	BTS → MS	Identification of the BTS (cell identity) and the location area and further information about organization of the CCCHs within the BTS. The Glossary provides an example of a BCCH/SYS_INFO 3.
1C	SYStem INFOrmation 4	BTS → MS	SYS_INFO 4 only repeats information of data already sent in the SYS_INFOs 1 - 3.
1D	SYStem INFOrmation 5	BTS → MS	The BTS uses SYS_INFO 5 (via SACCH) to inform the MS, during an active connection, about the BCCH frequencies of the available neighbor cells. This is particularly important after a handover when the MS cannot read the SYS_INFOs 1–4 of the new BTS.
1E	SYStem INFOrmation 6	BTS → MS	During an active connection, the current BTS (serving cell) provides the MS with all the necessary data of the serving cell by means of the SYS_INFO 6 (via SACCH).
1F	SYStem INFOrmation 7	BTS → MS	SYS_INFO 7 and 8 are used only for DCS1800 and PCS1900 to provide the registered MSs with additional information to access the serving cell (cell selection parameters).
21	PAGing REQuest Type 1	BTS → MS	Three different PAG_REQ messages were defined for activation of the MS in the case of an MTC. The difference between the messages lies simply in the number of MSs that can be paged simultaneously with one message (PAG_REQ 1 allows paging of two MSs, PAG_REQ 2 allows paging of three MSs, PAG_REQ 3 allows paging of four MSs). Consequently, the according number of IMSIs/TMSIs are contained in a PAG_REQ. Note that the IMSI is not contained in the PAG_REQ if a TMSI is assigned, even though the PAGING message on the A-interface contains both parameters.
22	PAGing REQuest Type 2	BTS → MS	
24	PAGing REQuest Type 3	BTS → MS	
27	PAGing ReSPonse	MS → BTS	PAG_RSP is the first message sent by the MS on the SDCCH to the BTS in an MTC. PAG_RSP corresponds to the CM_SERV_REQ message of a MOC.
28	HaNDover FAilure	MS → BTS	After an unsuccessful handover initiated by a HND_CMD, the MS sends a HND_FAI over the still existing connection to the old BTS.
29	ASSignment COMplete	MS → BTS	The MS confirms that it successfully changed to the (new) traffic channel, that is, the one previously assigned by an ASS_CMD message.

Table 7.5 (continued)

ID (Hex)	Name	Direction	Description
2B	HaNDover CoMmanD	BTS → MS	Channel assignment for a handover in which the BTS changes is always performed with HND_CMD; in an intra-BTS handover, the HND_CMD can be used. The message contains a description of the new traffic channel and the handover reference.
2C	HaNDover COMplete	MS → BTS	After successful handover initiated by a HND_CMD, the MS responds to the BTS with a HND_COM.
2D	PHYSical INFOrmation	BTS → MS	PHYS_INFO is the only message actually generated by the BTS. It is used in case of a nonsynchronized handover and is sent to the MS on the new channel $Ny1$ times. The content of the PHYS_INFO consists of the TA that the MS has to use initially.
2E	ASSignment CoMmanD	BTS → MS	Assignment of a traffic channel in case of an intracell handover or during call setup.
2F	ASSignment FAIlure	MS → BTS	The MS was not successful in changing to the channel specified in the ASS_CMD message. It has, therefore, changed back to the previously used channel and reports the failed access in a ASS_FAI message.
32	CIPHering MODe COMplete	MS → BTS	The MS confirms that a CIPH_MOD_CMD was received and that it has changed to the cipher mode.
35	CIPHering MODe CoMmanD	BTS → MS	The content of the CIPH_MOD_CMD message originates from the VLR. It is part of the ENCR_CMD message on the Abis-interface. The BTS informs the MS with CIPH_MOD_CMD that all data in both, uplink, and downlink are to be encrypted. The only content is the information as to which encryption algorithm A5/X shall be used.
39	IMMediate ASSignment EXTended	BTS → MS	The task of the IMM_ASS_EXT message is similar to that of the IMM_ASS_CMD message. The difference between the two is that the IMM_ASS_EXT message allows assignment of an SDCCH simultaneously for two MSs. That allows the network to reduce the number of messages. It is particularly helpful when the number of available AGCHs is low.
3A	IMMediate ASSignment REJect	BTS → MS	The BSC may answer a CHAN_REQ message with IMM_ASS_REJ if no SDCCHs are available. In this case, no channel is assigned and the MS is informed about a waiting period, during which it may not send a subsequent CHAN_REQ.

Table 7.5 (continued)

ID (Hex)	Name	Direction	Description
3B	ADDitional ASSignment	BTS → MS	There are some cases in which it may become necessary to assign a second halfrate traffic channel when one halfrate channel is already established, for example, to extend the bandwidth of the current connection for data transfer. In that case, the network sends to the MS an ADD_ASS message describing the new channel.
3F	IMMediate ASSignmnent CoMmanD	BTS → MS	The BSC uses the IMM_ASS_CMD to assign an SDCCH to the MS after a CHAN_REQ message was received. IMM_ASS_CMD is always sent on an AGCH. The message has to be distinguished from ASS_CMD, which is used to assign a traffic channel.

Table 7.6
Mobility Management (Skip Indicator/Protocol Discriminator = 05)

ID (Hex)	Name	Direction	Description
01/41	IMSI DETach INDication	MS → BTS	If IMSI attach/detach is allowed in the PLMN, then every time the MS is switched off the MS sends a IMSI_DET_IND to the MSC/VLR. This allows to more quickly reject an incoming call, or apply secondary call treatment, i.e., without sending PAG_REQ's first.
02	LOCation UPDating ACCept	BTS → MS	The MSC/VLR confirms a successful Location Update with a LOC_UPD_ACC. In some cases the LOC_UPD_ACC is used to assign a new TMSI as well.
04	LOCation UPDating REJect	BTS → MS	If a Location Update is not successful, (e.g., HLR is not reachable, IMSI or TMSI are unknown, etc.), then the MSC/VLR terminates the process with a LOC_UPD_REJ.
08/48	LOCation UPDating REQuest	MS → BTS	The MS sends the LOC_UPD_REQ to the MSC/VLR when it changes the Location Area, when Periodic Location Update is active, and when the MS is switched on again (with active IMSI attach/detach). LOC_UPD_REQ is part of the Location Update procedure.
11	AUTHentication REJect	BTS → MS	The AUTH_REJ message is used to inform the MS that authentication was not successful if the MSC/VLR found that the result for SRES from the MS was incorrect.

Table 7.6 (continued)

ID (Hex)	Name	Direction	Description
12	AUTHentication REQuest	BTS → MS	The MSC/VLR sends an AUTH_REQ message during connection setup, in order to authenticate the MS. The only parameter is RAND.
14/54	AUTHentication ReSPonse	MS → BTS	Answer to AUTH_REQ. It contains the authentication result SRES, which was determined by applying the values of K_i and RAND to the algorithm A3.
18	IDENTity REQest	BTS → MS	Although IDENT_REQ generally allows to request all three identification numbers from the MS, (IMSI, TMSI, and IMEI,) it is typically used by the Equipment Identity Register to request the IMEI only.
19/59	IDENTity ReSPonse	MS → BTS	IDENT_RSP is the answer to IDENT_REQ. The MS provides the network with the requested identification numbers (IMSI, TMSI, IMEI), which were requested in the IDENT_REQ message.
1A	TMSI REALlocation CoMmanD	BTS → MS	For every new connection, the VLR assigns a new TMSI to the MS in order to make tracking and interception of a subscriber more difficult. For this purpose, after the ciphering is active, the TMSI_REAL_CMD message is sent to the MS at any arbitrary position within the scenario.
1B/5B	TMSI REALlocation COMplete	MS → BTS	The MS confirms the receipt of a TMSI with a TMSI_REAL_COM.
21	CM SERVice ACCept	BTS → MS	Is used by the MSC if ciphering is not active or after the establishment of a second simultaneous CC connection. CM_SERV_ACC confirms to the MS that the service request, sent to the MSC in a CM_SERV_REQ message, was processed and accepted.
22	CM SERVice REJect	BTS → MS	The service request in which the MS has sent in a CM_SERV_REQ message is rejected by the MSC. The reason (e.g., overload) is provided.
23/63	CM SERVice ABOrt	MS → BTS	Is sent if a MS wants to terminate a MM connection. The CM_SERV_ABO can only be sent during a very narrow time window, because this message can only be used prior to the fist CC message sent.
24/64	CM SERVice REQuest	MS → BTS	The MS sends a CM_SERV_REQ at the beginning of every mobile originated connection in order to provide its identity (IMSI/TMSI) to the NSS, and to specify the service request in more detail (activation SS, MOC, Emergency Call, and SMS).

Table 7.6 (continued)

ID (Hex)	Name	Direction	Description
28/68	CM REeStablishment REQuest	MS → BTS	An option in GSM is to allow for a call reestablishment in case of a dropped connection. In these cases, first a CHAN_REQ has to be sent to the BTS and then it is tried with the CM_RES_REQ to reestablish an RR connection for the still existing and active MM and CC connection.
29	ABORT	BTS → MS	Is sent to the MS in order to release all MM connections. A possible reason is that the mobile equipment was identified as stolen (IMEI check). If this is actually the reason for sending ABORT, then the mobile equipment automatically blocks the Subscriber Identity Module. The SIM can, however, after switching off/on be used again.
31/71	MM STATUS	MS ↔ BTS	If one side receives a message for Mobility Management, which contains a protocol error in Layer 3, then an MM STATUS message with the respective error cause is sent. This kind of protocol error may be caused by bit errors on the Air-interface.

Table 7.7
Call Control (Transaction Identifier/Protocol Discriminator = X3)

ID (Hex)	Name	Direction	Description
01/41	ALERTing	MS ↔ BTS	The MSC sends this message in case of a Mobile Originating Call to the MS. In case of a Mobile Terminating Call, the MS sends an ALERT to the MSC. ALERT corresponds to the Address Complete Message (ACM) of ISUP and is responsible for the generation of a ring back tone at the receiving end. ALERT is always sent to that side of the call, which initiated it. This is important for protocol analysis.
02	CALL PROCeeding	BTS → MS	Is sent by the MSC in case of a Mobile Originating Call, in order to inform the MS that the address information which the MS has sent to the MSC in the SETUP message was received and processed. From the perspective of the MSC, CALL_PROC can be regarded as a confirmation that the ISUP Initial Address Message (IAM) was sent. The consequence for the MS is that the MSC does not need, or is not even able to process additional address information.

Table 7.7 (continued)

ID (Hex)	Name	Direction	Description
03	PROGRESS	BTS → MS	If, for a Mobile Originating Call, interworking or transport of inband signaling should become necessary, then the PROGRESS message is sent instead of ALERT. Examples are calls to automated information services or voice-mail boxes. In this case, the PROGRESS message can be regarded as a substitute for ALERT.
05/45	SETUP	MS ↔ BTS	When initiating a Mobile Originating Call, this message is sent by the MS to the MSC. The most important information are the address information of the called party and the type of connection, which is requested (Bearer Capabilities). In case of a Mobile Terminating Call, the MSC sends a SETUP message to the MS. When CLIP (Calling Line Identification Presentation) is active for the called party and is not restricted by the calling party, then the SETUP message also contains the directory number of the caller. The SETUP message is, furthermore, used to activate the Call Waiting tone (Supplementary Service) at the MS.
07/47	CONnect	MS ↔ BTS	The MSC sends this message during a Mobile Originating Call to the MS, to indicate that the connection was successfully established. The MS receiving the CON message corresponds to the MSC receiving the ISUP Answer Message (ANM). The MS sends a CON message to the MSC in case of a mobile terminanting call, as soon as the called party accepts the call.
08/48	CALL CONFirmed	MS → BTS	After receiving a SETUP message during a Mobile Terminating Call scenario, the MS confirms to the MSC in a CALL_CONF that it is able to establish the requested connection (Bearer Service, halfrate/fullrate, baud rate, etc.).
0E/4E	EMERGency SETUP	MS → BTS	This message is sent by the MS in case of an Emergency Call instead of a regular SETUP to carry address information.
0F/4F	CONnect ACKnowledge	MS ↔ BTS	CON_ACK is acknowledgment for a CON message. A call set up is regarded to be successful only after this message was sent. In particular charging starts typically with the CON_ACK message.

Table 7.7 (continued)

ID (Hex)	Name	Direction	Description
10/50	USER INFOrmation	MS ↔ BTS	It is possible in some cases to directly exchange data between the MS and its peer (e.g., ISDN or other MS). The maximum length of the transported payload is 128 octet, within GSM. For transport between GSM and some outside network, this maximum length may be restricted even further, depending on the capabilities of that other network (between 32 octet and 128 octet).
13/53	MODify REJect	MS ↔ BTS	MOD_REJ is the negative response to a MOD message. If the MS is unable to perform the adaptation which was requested by the peer, then the MS or the MSC respectively answers with a MOD_REJ. The reject cause is included in the message.
17/57	MODify	MS ↔ BTS	In some cases, it may become necessary to change the transmission parameters of an existing connection. This applies in particular, when a change from speech to data is made (Bearer Services 61 and 81). The MOD message carries out this task.
18/58	HOLD	MS → BTS	The HOLD message is used to put a call on hold when the user of a MS, while engaged in an active call, receives a second incoming call or wants to set up another call (Multiparty). Then the HOLD message is sent to the MSC. Hold is also the name of the related Supplementary Service.
19	HOLD ACKnowledge	BTS → MS	Acknowledgment by the MSC that a call was placed in the hold state after a HOLD message was received.
1A	HOLD REJect	BTS → MS	The MSC was unable to place a call into hold state. Therefore, the HOLD message is answered with a HOLD_REJ. The reason for this rejection is given in the cause value.
1C/5C	RETRIEVE	MS → BTS	The MS sends this message in order to reactivate a connection which was previously placed on hold.
1D	RETRIEVE ACKnowledge	BTS → MS	The MSC confirms that it has received and processed the RETRIEVE message. The call which was placed on hold is now active again.
1E	RETRIEVE REJect	BTS → MS	It is not possible to switch back to a call that was put on hold. The RETRIEVE request gets a negative response.

Table 7.7 (continued)

ID (Hex)	Name	Direction	Description
1F/5F	MODify COMplete	MS ↔ BTS	MOD_COM is the acknowledgment of a MOD message. Depending on the direction, MOD_COM is sent at different points in a scenario. The MSC sends MOD_COM only after the requested adaptation has been performed. The MSC sends this message already after receiving and accepting the MOD message.
25/65	DISConnect	MS ↔ BTS	Is used either by the MSC or the MS, to terminate an existing CC connection. The DISC message always contains a cause value, which indicates the reason why the connection was disconnected. When the call is terminated regularly, the cause value "16" is sent, which stands for 'normal clear'. Another value in case of problems is e.g., cause 47 = Resources unavailable. Please be advised that when analyzing trace files, even in case of errors the DISC message may carry a normal clear. This is the case when the problem was not detected by call control.
2A/6A	RELease COMplete	MS ↔ BTS	REL_COM is the answer to a REL message and the acknowledgment that the CC resources have been released. REL_COM is always sent by that side, which had previously sent the DISC message. Like for REL, also for REL_COM there exists an ISUP message with the same name.
2D/6D	RELease	MS ↔ BTS	Because of the fact that signaling in GSM is related to ISDN, there are some similarities in the CC protocol between the two. The REL message corresponds directly to an ISUP message with the same name, which in the case of ISDN is responsible for terminating a connection. The same functionality provides this message in GSM, namely to release the CC resources. The relationship is illustrated in Chapter 12, "Scenarios".
31/71	STOP DTMF	MS → BTS	It is possible to use DTMF signaling with a MS. For this purpose, a START_DTMF message is sent to the MSC when the user presses a button on the keypad. This tells the MSC to generate the respective DTMF sound and send it inband to the peer entity (ISDN, PSTN) When the user releases the button, a STOP_DTMF message is sent to the MSC which triggers the MSC to stop sending the respective tone.

Table 7.7 (continued)

ID (Hex)	Name	Direction	Description
32	STOP DTMF ACKnowledge	BTS → MS	Acknowledgment by the MSC that a STOP_DTMF message was received and sending of the DTMF tone was stopped.
34/74	STATUS ENQuiry	MS ↔ BTS	Both MS and MSC may use STATUS_ENQ to inquire about the current state of Call Control in the peer entity. The peer has to answer the STATUS_ENQ with a STATUS message otherwise the connection is torn down.
35/75	START DTMF	MS → BTS	The MS uses START_DTMF to send ASCII coded DTMF tones to the MSC. The only content of a START_DTMF message is the ASCII value of the respective button, which was pressed at the MS. This is for example 31hex when the '1' button was pressed. A START DTMF message can only be sent in a traffic channel during an active connection. Please note that it is not possible to transmit an analog DTMF tone between the MS and the MSC, only the START_DTMF message. The tone, which can be heard at the MS at the same time is generated in the MS. The Glossary provides a detailed description on the transmission of DTMF tones.
36	START DTMF ACKnowledge	BTS → MS	START_DTMF_ACK is the acknowledgment of the MSC that a START_DTMF message was received. When the MSC sends the START_DTMF_ACK, it simultaneously sends an analog DTMF tone which is sent inband in a traffic channel towards the PSTN/ISDN. The duration of the tone is determined by when a STOP_DTMF message is received.
37	START DTMF REJect	BTS → MS	When the MSC is unable to process the START_DTMF, then it sends a START_DTMF_REJ message to the MS. The respective reason is included in the cause value.
39/79	CONGESTion CONTROL	MS ↔ BTS	This message may be used by both sides, in order to activate flow control for data which is transported within USER_INFO messages.
3D/7D	STATUS	MS ↔ BTS	A STATUS message can be sent if protocol errors in the area of Call Control are detected or if a STATUS_ENQ has to be answered. Such an error situation can occur, in particular, when misinterpretations of CC messages occur, because of bit errors (refer also to MM_STATUS and RR_STATUS).

Table 7.7 (continued)

ID (Hex)	Name	Direction	Description
3E/7E	NOTIFY	MS ↔ BTS	The NOTIFY message is used in case of an active connection to inform the peer entity about a incident in the area of Call Control. Example: When a GSM subscriber is placed on Hold because the other party intends to accept or establish another call, then the MSC sends a NOTIFY message to that MS.

Table 7.8
Supplementary Services (Transaction Identifier/Protocol Discriminator = XB)

ID (Hex)	Name	Direction	Description
2A/6A	RELease COMplete	MS ↔ BTS	Although already presented for the Call Control, the REL_COM message shall be separately presented for Supplementary Services. If a connection was established because of a Supplementary Services request, then this connection is released by sending a REL_COM message. [GSM 04.10, GSM 04.80]
3A/7A	FACILITY	MS ↔ BTS	The FACILITY message may be used by both the MS as well as the NSS. The content of this message is transparent data for Supplementary Services. Please note that almost all CC messages contain an optional information element, the 'Facility', with which SS information can be transported without requiring a FACILITY-message. [GSM 04.10, GSM 04.80]
3B/7B	REGISTER	MS ↔ BTS	The REGISTER message is needed for the activation or inquiry of call-independent supplementary services. Example: the activation of Call Forwarding. In this case, sending of a REGISTER message implies that a new Transaction Identifier is assigned and the dialog between MS and the network is established. [GSM 04.10, GSM 04.80]

8

Signaling System Number 7

Signaling System Number 7 (SS7) provides in OSI Layers 1 to 3 the basis for the signaling traffic on all NSS interfaces, as well as on the A-interface.

The relation to GSM is rather hidden at the beginning of the description, but it becomes more and more obvious when the various user parts like SCCP and TCAP/MAP are presented. SS7, together with all its functionality and user parts, forms a much more complex signaling system than LAPD and $LAPD_m$. For that reason, this whole chapter is dedicated to SS7, or rather to its Layers 1 to 3.

8.1 The SS7 Network

A SS7 network consists of directly connected signaling points (SPs), as shown in Figure 8.1(a); SPs that are connected through signaling transfer points (STPs), as shown in Figure 8.1(b); or a combination of SPs and STPs, as shown in Figure 8.1(c). An SP is a network node that has user parts (e.g., SCCP, ISUP) that allow the processing of messages addressed to that SP. The MSC, the BSC, and the exchanges of the PSTN fall into this category. The functionality of the STP typically is related to those of the SP, with the additional capability of being able to relay SS7 messages. Note that it is possible to have a designated STP that has no SP functionality, that is, one that can only relay messages, as shown in Figure 8.1(b).

Figure 8.1(a) Directly connected SPs.

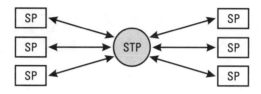

Figure 8.1(b) An STP that interconnects SPs.

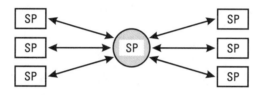

Figure 8.1(c) An STP with SP functionality that interconnects SPs.

8.2 Message Transfer Part

SS7 without user parts consists only of the OSI Layers 1 through 3. Those three layers essentially are represented by the message transfer part (MTP). Parts of the SCCP actually are also part of Layer 3.

The MTP of SS7 performs the following general tasks:

- It provides all the functionality of OSI Layers 1 to 3 required to provide for a reliable transport of signaling data to the various SS7 user parts.
- When problems arise, the MTP takes the necessary measures to ensure that the connection can be maintained or prevents loss of data, for example, by switching to an alternative route.

The MTP can be partitioned into three layers, where the MTP 1 (OSI Layer 1) is responsible for the transfer of single bits or the definition and provision of the necessary electrical and physical means for it.

The MTP 2 (OSI Layer 2) defines the basic frame structure that is used by SS7 for all message types. This frame structure is illustrated in Figure 8.2, which shows the flags that mark beginning and end (as we have seen in LAPD),

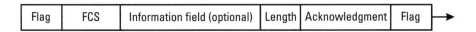

Figure 8.2 General format of an SS7 message.

an acknowledgment field with send and receive sequence numbers, a length indicator, an optional information field, and the FCS to transport a checksum (as we also have seen in LAPD).

8.3 Message Types in SS7

Definition of the SS7 message types is another functionality of MTP 2. In Layer 2 of SS7, three different message types are defined: the fill-in signal unit (FISU), the link status signal unit (LSSU), and the message signal unit (MSU). Although no explicit field is available to distinguish among the message types, it is possible to do so based on their different lengths. The length indication (LI) provides that information and relates to the length of the optional data field. The value of the LI is always 0 for FISUs; 1 or 2 for LISUs; and greater than 2 for MSUs.

8.3.1 Fill-In Signal Unit

The FISU (Figure 8.3) is used to supervise the link status when no traffic is available. Both sides poll each other in this idle state. The N(S) and N(R), which in SS7 are called the forward sequence number (FSN) and backward sequence number (BSN) or the forward indicator bit (FIB) and backward indicator bit (BIB), respectively, do not change their values during polling. In addition to the polling functionality, an FISU also can be used to acknowledge receipt of an MSU.

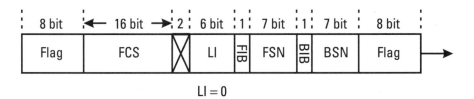

Figure 8.3 Format of the FISU.

8.3.2 Link Status Signal Unit

The LSSU is used only to bring a link into service or to take it out of service and during error situations (e.g., overload), to exchange status information between two SPs or STPs. In Figure 8.4, the status field has a length of 1 byte, but according to ITU definitions it also can be 2 octets long. In any case, only the first three bits of the first byte contain the actual status information. The receiver of an LSSU does not confirm its receipt.

Protocol test equipment usually does not indicate an LSSU as such but displays it according to its status field. For that reason, the status field or its abbreviation can also be used as a subname. Consequently, the term status indication (SI) and the terms SIO, SIOS, and SIB, which are explained in Table 8.1, are used more frequently than LSSU/status field SIOS. Note that, in particular, when SIPOs or SIBs are detected during protocol testing, rather serious problems can be expected at the related SP/STP.

8.3.3 Message Signal Unit

The MSU (Figure 8.5) is used for any type of data transfer between two network nodes. The MSU is the only SS7 message able to carry traffic data (the LSSU does not carry traffic data, only status information), and it is used by all user parts (ISUP, SCCP, OMAP) as a platform particularly for that task. The information field of the MSU consists of the service information octet (SIO)

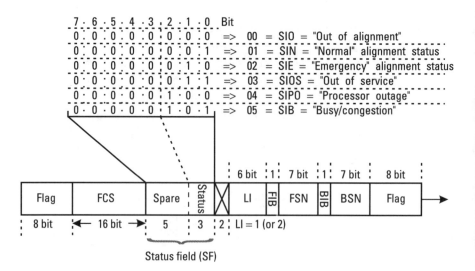

Figure 8.4 Format of the LSSU.

Table 8.1
Status Field Values

Value	Abbreviation	Status	Description
0	SIO	Out of alignment	Start of link alignment
1	SIN	Normal alignment	A connection is brought into service with a normal (long) surveillance time of 8.2 s (see also Section 8.4.2).
2	SIE	Emergency alignment	A connection is brought into service with an emergency (short) surveillance time of about 500 ms. (see also Section 8.4.2).
3	SIOS	Out of service	This indicates in case of an error situation or before a link is taken into service that currently no MSUs can be sent or received.
4	SIPO	Processor outage	When the Layer 2 of an SP or STP detects a problem within the Layer 3 of its own network node it indicates the problem status to the peer entity by sending a SIPO.
5	SIB	Busy/congestion	Signals overload on the originating side. Acknowledgments cannot be sent anymore. Usually, a link failure follows.

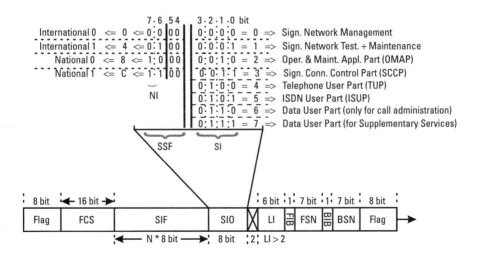

Figure 8.5 Format of the MSU.

and the signaling information field (SIF). The SIO[1] is further partitioned into the subservice field (SSF) and the service indicator (SI), with four bits each. Only the two bits of higher valence in the SSF are necessary to describe the network indicator (NI). The NI is used to distinguish between national and international messages. The SI indicates to which user part an MSU belongs. The four bits of the SI thus determine whether the data of the SIF belong to the SCCP, TCAP, ISUP, and so on, or possibly need to be forwarded to the automatic network management. In contrast to LSSU and FISU, it has to be acknowledged to the peer entity whenever an MSU is received.

8.4 Addressing and Routing of Messages

In an SS7 network, MSUs are not necessarily exchanged between adjacent neighbors (SP/STP). In a GSM system, the MSC and BSC are neighbors; however, the exchange of information between the MSC and the HLR may involve several STPs. SS7 uses so-called point codes for routing and addressing MSUs. Point codes are unique identifiers within an SS7 network. Exactly one point code, a signaling point code (SPC), is assigned to every SP and STP. An MSU has a routing label that contains the point codes of the sender (the originating point code, or OPCs) and the addressee (the destination point code, or DPC). The routing label is, for its part, a component of the SIF. Note that neither FISU nor LSSU possesses a routing label, since those messages are exchanged only between two adjacent nodes.

Figure 8.6 shows the format of a routing label. The OPC defines the sender of the MSU, and the DPC defines its addressee. Note that addressing via SPCs works only on a national basis. The services of higher layers are needed for international addressing, in particular SCCP or ISUP, to provide the necessary features.

The remaining 4 bits of the routing label form the signaling link selection (SLS) field. This parameter is used to balance the load between several SS7 connections of a link group. If, for example, two SS7 connections are available between two network elements, all the even values of the SLS $(0, 2, 4,\ldots, 14_{dez})$ are assigned to the first link, and the odd values $(1, 3, 5,\ldots, 15_{dez})$ are assigned to the second link. This fact is important to know in the analysis of SS7 trace files.

---

1. Note that the service information octet and its abbreviation SIO do not have a relation to the former use of the abbreviation, which stood for status indication/out of alignment. Unfortunately, the standards use the same abbreviation for both.

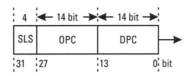

Figure 8.6 Routing label (DPC, OPC, and SLS).

8.4.1 Example: Determination of DPC, OPC, and SLS in a Hexadecimal Trace

In the analysis of hexadecimal trace files, it generally is important to be able to convert DPC and OPC into clear text, to be able to relate the various messages to, for example, a particular MSC or BSC. As shown in Figure 8.6, DPC and OPC are each 14 bits long. The routing label, together with the 4 bits of the SLS, totals 32 bits, or 4 bytes.

Because the OPC and the DPC are 14 bits in length, it is not trivial, particularly with byte (8 bits) or 16-bit-word-oriented presentations, to derive the decimal value of DPC or OPC, as illustrated in Figure 8.7. The sequence of numbers represents the hexadecimal values. The underlined part represents the routing label, that is, the SLS, OPC, and DPC. This information is decoded in clear text. At first sight, the values seem to differ.

It is important when decoding to consider the bitwise sequence of transmission with which the data are received by the system. The binary presentation (left to right) is given in Figure 8.8.

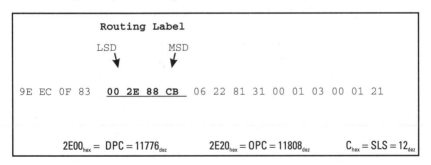

Figure 8.7 Partial trace file and point codes.

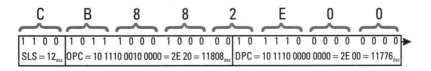

Figure 8.8 Transmission of routing label.

Possible confusion is based on the unusual length (14 bits) of OPC and DPC on the one hand, and, on the other hand, the results from the reversed way of reading/writing (right to left), a problem familiar to most programmers. Misinterpretation can be prevented when these facts are considered.

Other representations of SPCs can be used in various national applications, like the "4-3-4-3" presentation, which refers to the bits that are used per sign. The example in Figure 8.7 reads in the "4-3-4-3" presentation as follows:

$$DPC = 2E00_{hex} = 11776_{dez} = 1011 - 100 - 0000 - 000 = B - 4 - 0 - 0$$

$$OPC = 2E20_{hex} = 11808_{dez} = 1011 - 100 - 0100 - 000 = B - 4 - 4 - 0$$

8.4.2 Example: Commissioning of an SS7 Connection

Every SS7 connection is brought into service as presented in Figure 8.9. In the figure, an A-interface link between BSC and MSC is brought into service.

8.4.2.1 Bringing Layer 2 Into Service

After Layer 1 is established, both sides send an SIOS-LSSU, which indicates that the link is out of service and no MSU can currently be processed.

The process to bring Layer 2 into service starts with sending an SIO-LSSU. Please note the duplex characteristics of SS7. Both terminals are equal, and a link has to be established in *both* directions.

The test period, during which both sides examine the link quality, starts with sending an LSSU-SIN or an LSSU-SIE. Transmitted FISUs must not contain any errors during this test period. The link cannot go into service if an error occurs. The difference between LSSU-SIE and LSSU-SIN is the related surveillance time.

An emergency alignment is used when no alternative SS7 route currently exists and the link needs to be in service as quickly as possible.

8.4.2.2 Bringing Layer 3 Into Service

When the test time is over and no errors were detected, Layer 2 is considered to be in service and Layer 3 initiates further tests. A signaling link test message (SLTM) is used for that purpose, to transmit a number of test bytes to Layer 3 of the peer entity.

If the test bytes are correctly returned to the sender in a signaling link test acknowledgment (SLTA) message, Layer 3 is also considered to be "in traffic." Figure 8.10 shows examples of a SLTM and a SLTA message.

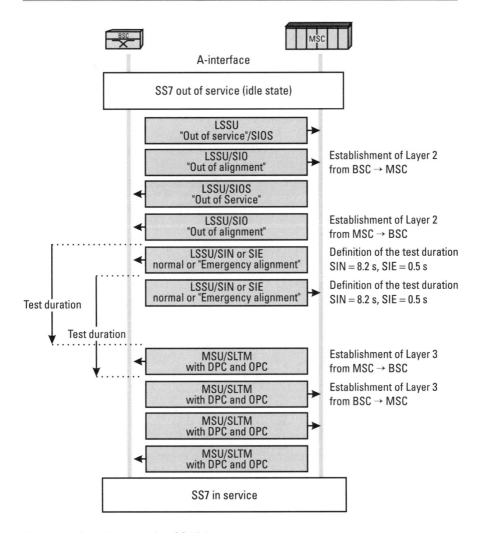

Figure 8.9 Establishment of an SS7 link.

Synchronization of Layer 4 (in this case of the SCCP) follows the link establishment on the A-interface, by applying the reset procedure (described in Chapter 10).

8.5 Error Detection and Error Correction

Layer 2 is responsible for error detection and error correction. To be more specific, within Layer 2, the FSN and the BSN, together with the FCS, take care

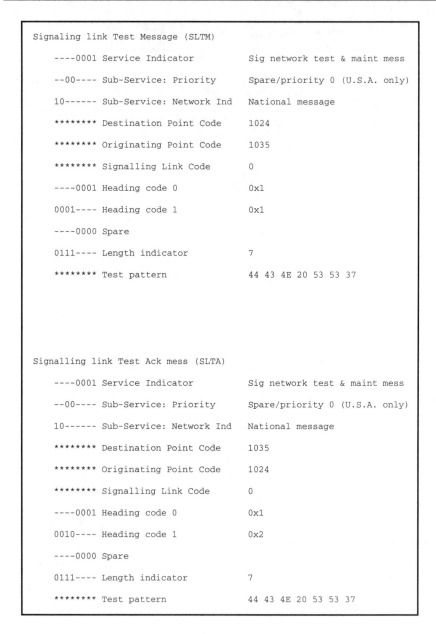

Figure 8.10 Examples of a SLTM and a SLTA message.

of the error recognition function. Note that the format of those parameters is the same for all three message types (FISU, LSSU, MSU). Refer to Figures 8.3 through 8.5.

SS7 provides two alternative methods of error correction:

- All messages not acknowledged within a specified time frame have to be retransmitted.
- Retransmission occurs only in the case of a negative acknowledgment.

These two basic procedures are described next.

8.5.1 Send Sequence Numbers and Receive Sequence Numbers (FSN, BSN, BIB, FIB)

The 7-bit-long FSN, together with the FIB, forms the send sequence number of an SS7 message. The FSN is incremented by 1 whenever an MSU is sent. The value of the FSN, however, does not change when an LSSU or an FISU is transmitted.

The 7-bit-long BSN and the BIB form the receive sequence number. They are used for positive or negative acknowledgment of a received message. The BIB is used to indicate a problem when a negative acknowledgment has to be returned because of a transmission error. To indicate a transmission error, the value of BIB is simply inverted by the receiving entity, that is, changed from 0 to 1 or from 1 to 0. The inverted value of BIB is sent back to the peer entity together with the BSN of the last error-free received MSU. The peer entity then has to repeat all MSUs with a greater BSN.

FSN/FIB on one side and BSN/BIB on the other together form a functional unit, as Figure 8.11 shows. The example in Section 8.5.2 describes the task of FSN, FIB, BSN, and BIB in more detail.

8.5.2 BSN/BIB and FSN/FIB for Message Transfer

The task of FSN and BSN can best be explained with an example. Refer to Figure 8.11, which shows the exchange of SS7 messages between BSC and MSC. The numbers in the following bulleted list relate to the line numbers in Figure 8.11. The intention is to explain the values of the counters and the principle of error correction.

- Line 1. Let us assume, for simplification purposes, that the link was just brought into service and that the values for FIB, FSN, BIB, and BSN are all 1.
- Line 2. The BSC increments its FSN to a value of 2 and sends an MSU. The other counters do not change. The BSC continues to store

Nr.	BSC FIB	BSC FSN	BSC BIB	BSC BSN	Message	MSC FIB	MSC FSN	MSC BIB	MSC BSN
1	1	1	1	1		1	1	1	1
2	1	1 2	1	1	MSU/FIB = 1 / FSN = 2/BIB = 1/BSN = 1 (BSC → MSC)	1	1	1	1→2
3	1	2	1	1→2	MSU/FIB = 1/FSN = 2/BIB = 1/BSN = 2 (MSC → BSC)	1	1→2	1	2
4	1	2	1	2→3	MSU/FIB = 1/FSN = 3/BIB = 1/BSN = 2 (MSC → BSC)	1	2→3	1	2
5	1	2	1	3→4	MSU/FIB = 1/FSN = 4/BIB = 1/BSN = 2 (MSC → BSC)	1	3→4	1	2
6	1	2	1	4	FISU/FIB = 1/FSN = 2/BIB = 1/BSN = 4 (BSC → MSC)	1	4	1	2
7	1	2→3	1	4	MSU/FIB = 1/FSN = 3/BIB = 1/BSN = 4 (BSC → MSC)	1	4	1	2→3
8	1	3→4	1	4	MSU/FIB = 1/FSN = 4/BIB = 1/BSN = 4 (BSC → MSC)	1	4	1	3→4
9	1	4→5	1	4	MSU/FIB = 1/FSN = 5/BIB = 1/BSN = 4 (BSC → MSC)	1	4	1	4→5
10	1	5→6	1	4	MSU/FIB = 1/FSN = 6/BIB = 1/BSN = 4 (BSC → MSC)	1	4	1	5
11	1	6→7	1	4	MSU/FIB = 1/FSN = 7/BIB = 1/BSN = 4 (BSC → MSC)	1	4	1	5
12	1 0	7→5	1	4	FISU/FIB = 1/FSN = 4/BIB = 0/BSN = 5 (MSC → BSC)	1	4	1→0	5
13	0	5→6	1	4	MSU/FIB = 0/FSN = 6/BIB = 1/BSN = 4 (BSC ↔ MSC)	1	4	0	5→6
14	0	6→7	1	4	MSU/FIB = 0/FSN = 7/BIB = 1/BSN = 4 (BSC → MSC)	1	4	0	6→7
15	0	7	1	4	FISU/FIB = 1/FSN = 4/BIB = 0/BSN = 7 (MSC → BSC)	1	4	0	7

Figure 8.11 FSN, FIB and BSN, BIB for error correction.

the contents of the MSU for possible retransmission. The MSC receives the message and checks for errors. When the MSC finds that the message is correct, it increments its BSN counter from 1 to 2.

- Line 3. It happens that the MSC also has to send an MSU. Now the MSC increments its FSN counter from 1 to 2 and transmits this value, together with the new value of BSN (2) to the BSC. The MSC also continues to store the contents of the MSU. After receiving the message, the BSC checks the values for BSN and BIB and finds that the MSC has confirmed the previously sent (under line 2) MSU. The information is contained in the parameter BSN (BSN = 2). Since the MSC received the message without errors, the BSC does not need to continue to store that information and discards it. And because the BSC has received the message from the MSC without error, it increments the value of BSN to 2.

- Line 4. Now the MSC sends another MSU before the BSC is able to acknowledge the MSU. The value of FSN in the MSC increases accordingly to a value of 3, and the value of BSN in the BSC is changed to 3. In addition to the message from line 3, the new MSU has to be stored in the MSC.

- Line 5. The process described under line 4 repeats. The MSC now has to store all three unacknowledged MSUs.

- Line 6. Now the BSC acknowledges that it received the three messages (lines 3, 4, and 5) from the MSC without error. Note the value of BSN (BSN = 4) in the FISU. All three MSUs are acknowledged in one FISU message by confirming the latest correctly received message. Hence, it is not necessary to acknowledge every single message.

- Lines 7, 8, and 9. The BSC transmits three consecutive MSUs to the MSC. It correspondingly increases its value for FSN from 2 to 5. The MSC increments its value for BSN to 5 as well. The BSC needs to store all three MSUs until the MSC confirms proper receipt of them.

- Line 10. The BSC sends another MSU to the MSC, which increases the value of FSN in the BSC to 6. This MSU is corrupted and the MSC detects the error (FCS). Consequently, the value of BSN in the MSC does not change.

- Line 11. Now the BSC sends another MSU to the MSC before the MSC is able to send a negative acknowledgment. Although this message is received without error, the counter for BSN still is not incremented, and its value stays at 6.

- Line 12. Because of the error that occurred in line 10, the MSC inverts its value for BIB from 1 to 0. The new BIB value is sent back to the BSC in an FISU, together with the value for the number of the latest correctly received MSU. When the BSC analyzes this message, it detects from the BIB value that a transmission error occurred and that the MSUs sent under lines 7 through 9 are confirmed. The BSC then inverts its value for FIB and changes its value for FSN from 7 to 5, to retransmit the messages from lines 10 and 11. Note that the message, sent under line 11 and correctly received by the MSC, still has to be sent again.
- Lines 13, 14. With the inverted values of FIB and BIB on the respective sides, the BSC repeats transmission of the messages from lines 10 and 11 to the MSC. This time, both messages are received without error, and the value for BSN in the MSC is increased from 5 to 7.
- Line 15. The MSC confirms receipt of the MSUs (lines 13 and 14, previously lines 10 and 11) by answering with an FISU.

The preceding example can be summarized as follows:

- The counters for BSN/BIB on one hand and FSN/FIB on the other have identical values after positive acknowledgment.
- A corrupted message is indicated by the inversion of the BIB. This procedure also is used in the idle case, when only FISUs are exchanged.
- When one side receives a message with an inverted BIB, it subsequently inverts its FIB and sends it with the next message, to indicate to the peer that it has received the information about the error situation.
- When the values of the counters FSN, FIB, BSN, and BIB are not consistent in a received message (e.g., a jump from 1 to 5), a negative acknowledgment also is returned.
- The counters are not incremented when an FISU or an LSSU is sent.

8.6 SS7 Network Management and Network Test

The various possibilities to configure an SS7 network were presented at the beginning of this chapter. Such a network basically consists of interconnected SPs and STPs, as illustrated in Figure 8.12.

The major task in the operation of a complex network is management or administration of the network. For that purpose, SS7 has dedicated user parts

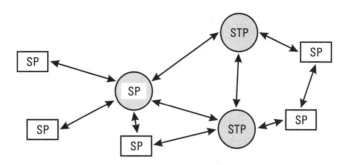

Figure 8.12 Example of an SS7 network.

in Layer 3 that automatically detect error situations and are able to try to correct them autonomously. Errors can be separated into one of three categories:

- Overload on a single SS7 link;
- Outage/bringing into service an SP/STP;
- Outage/bringing into service an SS7 link between SPs or STPs.

SS7 network management has to be able to detect each error situation and to allow for a fix with the appropriate actions. Those actions consist mainly in rerouting signaling data, as well as blocking and unblocking routes. The details of how that is performed are described next.

8.6.1 SS7 Network Test

ITU recommendations deal separately with SS7 network test and SS7 network management. However, they serve similar tasks and thus will be presented together. Table 8.2 shows how SS7 distinguishes network test and network management by assigning two different SIs. The SI is part of the SIO.

Table 8.2
Service Indicators for SS7 Network Management and Network Test

SI (Binary)	User Part
00	Network management
01	Test and maintenance

8.6.2 Possible Error Cases

The messages in the following descriptions frequently are abbreviated. Sections 8.6.5 and 8.6.6 explain the abbreviations

8.6.2.1 Behavior in an Overload Situation

Figure 8.13 illustrates a situation in which an overload situation occurs on the link between the STP and the SP/STP. In that case, both STPs inform their direct neighbors about the limited availability of that connection. The information is sent in TFC messages and TFR messages. Alternative routes will be used after the neighbors are informed about the problem. The changeover procedure (sending of a COO message) is used to actually implement rerouting. When the overload situation has ceased to exist, the neighbors are informed in TFA messages that the link is available again. To actually utilize that link, the changeback procedure has to be executed (sending of a CBD message). In the time period between when the TRC/TFR messages and the TFA messages are sent, the neighbor SPs/STPs periodically check the overload situation by sending RSR or RCT messages. Which of the messages is actually used depends on whether the sender is an SP or an STP. (See the tables at the end of the chapter for explanations of the messages.)

8.6.2.2 Behavior at Outage/Bringing SP/STP Into Service

Figure 8.14 shows the same SS7 network, but this time with the failure of an STP. As with the overload situation, all neighbors have to be informed immediately about the problem. In the case of an outage, that is performed by the sending of TFP messages to all affected SPs and STPs. The rerouting process, as in the overload situation, is done via the changeover procedure. When the failed STP comes back into service, the neighbors first recognize that when the links toward that STP synchronize again. All the neighbor nodes periodically send RST messages during the outage to check the state of the STP. All

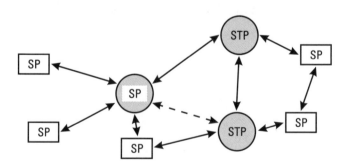

Figure 8.13 SS7 network with overload (dashed link).

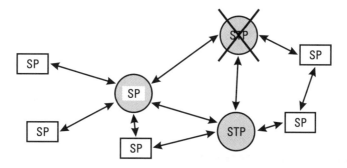

Figure 8.14 SS7 network with STP outage.

the neighbors send TRA messages to the STP when Layer 2 is established again. The STP that got back into service also sends TRA messages to all neighbors (an SP would *not* send those messages). The TRA messages have the purpose of informing all neighbors that the respective routes are available again. Finally, a changeback procedure is used to cancel all the established detours.

8.6.2.3 Behavior When SS7 Link Fails/Is Established

Another possible scenario is the total loss of an SS7 link between an SP and an STP (Figure 8.15). In that situation, the STP has to inform all neighbors about the loss of the connection. That is done with a TFP message. Establishing alternative routes to and from the affected SP is performed by means of the changeover procedure. Both the SP and the STP periodically test the link by sending RST messages during the time period when the link is not available.

The reverse process is executed when the link is brought back into service. The STP sends TFA messages to all affected neighbors, and the alternative routes are canceled by means of the changeback procedure.

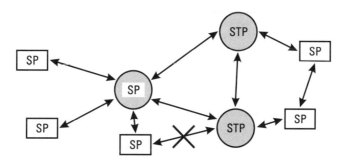

Figure 8.15 SS7 network in which a link has failed.

8.6.2.4 More Error Cases

Other situations may occur, and appropriate measures have to be taken. Such situations include loss of single user parts in an SP, intentional shutdown of an SS7 link by network management, and automatic addition of SS7 links in the case of increasing load or link failures. The tables in the next sections describe the corresponding messages to transmit that information.

8.6.3 Format of SS7 Management Messages and Test Messages

Figure 8.16 presents the format of SS7 management messages and test messages.

The SI is used to differentiate between the user parts for network management and network test.

The SI field of management messages and test messages (both in Layer 3) contains so-called heading codes, which are necessary for message and message-group coding. The SS7 user parts of Layer 4 do not allow for such a functionality. Heading code 0 (H0) defines a whole group, while Heading code 1 (H1) is used to identify a single message within a group. The defined message groups are listed in Table 8.3.

The data part (see Figure 8.16) is optional and not required by all messages.

8.6.4 Messages in SS7 Network Management and Network Test

Figures 8.17(a) through 8.17(d) provide a complete overview of the SS7 messages used for network management and network test. Tables 8.4 and 8.5 list

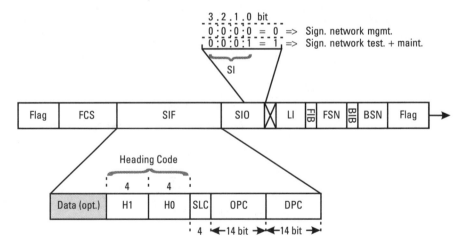

Figure 8.16 Format of SS7 management and test messages.

Table 8.3
Various Message Groups of SS7 Management and Test

H0 (Hex)	Message Group
1	Changeover and changeback
2	Emergency changeover
3	Transfer controlled and signaling route test
4	Transfer prohibited/allowed/restricted
5	Signaling route set test
6	Management inhibit
7	Traffic restart allowed
8	Signaling data link connection
A	User part flow control
1 (test)	Signaling link test

all SS7 network management and test messages, with brief descriptions thereof. Uppercase letters indicate the abbreviations used in this context.

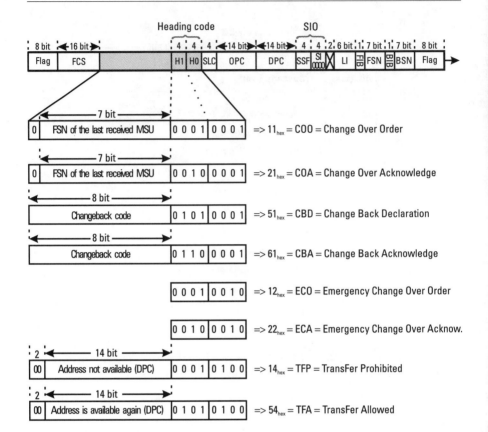

Figure 8.17(a) SS7 network management messages.

Signaling System Number 7

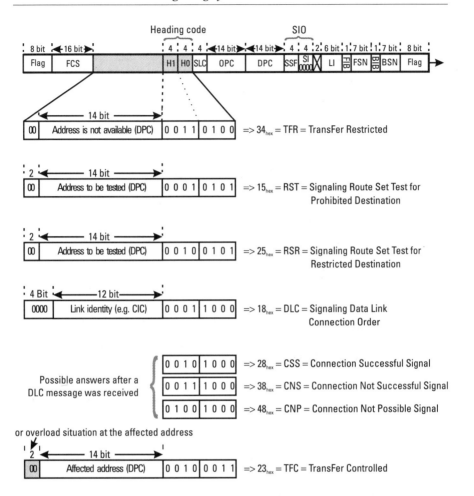

Figure 8.17(b) SS7 network management messages.

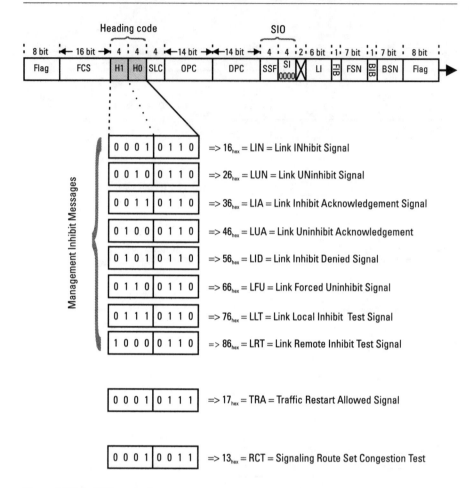

Figure 8.17(c) SS7 network management messages.

Signaling System Number 7

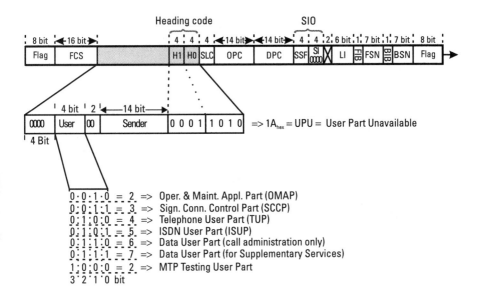

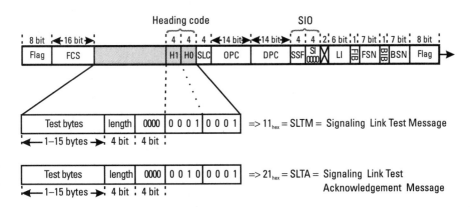

Figure 8.17(d) SS7 Management and test messages.

Table 8.4
Message Types in SS7 Network Management

H1/H0	Abbr.	Name	Description
11	COO	Change Over Order	The COO message is used to re-route the signaling traffic from one link to an alternative link. This is used when a particular signaling link cannot be used anymore. Possible reasons are, intentional shut down, or automatic network management procedures. The COO message always contains the FSN of the last MSU that was received and acknowledged on that link. This allows it to immediately continue with transmission over an alternative link, without loss of data. Both SP or STP can send a COO message.
21	COA	Change Over Acknowledge	The COA message confirms that a COO message was received and processed and that a particular signaling link will no longer be used. A COA message also acknowledges the FSN of the COO message.
51	CBD	Change Back Declaration	The CBD message indicates to the neighbor SP/STP that a previously shut down link, performed by the change over procedure, is now available again and can be used for signaling traffic.
61	CBA	Change Back Acknowledge	Answer to a CBD message. A previously shut down link is now available again and can be used for signaling traffic.
12	ECO	Emergency Change Over Order	It is possible during change over that the FSN of the last correctly received MSU can not be transmitted (e.g., Error situation). In such circumstances the ECO message is used instead of the COO message. The disadvantage of the emergency change over is that it is not possible to determine whether MSU's were lost.
22	ECA	Emergency Change Over Acknowledge	This message acknowledges an ECO message and the switch over to the alternative link(s).
14	TFP	TransFer Prohibited	The TFP message is only used by STPs to inform neighbor SPs or STPs that a certain route to a particular Point Code is not available. The TFP message contains as a parameter the Point Code to which no data can be sent. When a TFP message was received, all the traffic to the affected Point Code has to be routed via alternative routes.
54	TFA	TransFer Allowed	The TFA message is only sent by STPs in order to inform neighbor SPs/STPs about the availability of a route. This message is the counterpart of the TFP message and the TFR message, respectively.

Table 8.4 (continued)

H1/H0	Abbr.	Name	Description
34	TFR	TransFer Restricted	The TFR message is only sent by STPs in order to inform neighboring SPs/STPs about the limited availability of a route to a particular Point Code. Data destined to this Point Code should, when possible, be routed via alternative links. The TFR message is a softer alternative to the TFP message and is used, for example, in the case of overload.
15	RST	Signaling Route Set Test for Prohibited Destination	When an SP/STP receives a TFP message, it periodically sends a RST message to the affected STP, in order to determine if it is available again.
25	RSR	Signaling Route Set Test for Restricted Destination	When an SP/STP receives a TFR message, it periodically sends a RSR message to the affected STP, in order to determine if it is available again.
18	DLC	Signaling Data Link Connection Order	When alternative signaling connections need to be established between SPs/STPs, then a DLC message is sent to the neighbor SP/STP. The most important parameter is the identifier for the channel to be reserved (time slot).
28	CSS	Connection Successful Signal	A positive response when a DLC message is received. The requested channel is available for SS7 and all necessary resources were activated. The generic term for CSS, CNS, and CNP is Signaling Data Link Connection Acknowledgment.
38	CNS	Connection Not Successful Signal	A negative response when a DLC message is received. The requested channel is not available for SS7. The CNS message signals that further attempts to activate such a link can be made, which differentiates it from the CNP message. The generic term for CSS, CNS, and CNP is Signaling Data Link Connection Acknowledgement.
48	CNP	Connection Not Possible Signal	A negative response when a DLC message is received. The requested channel is not available for SS7. The CNP message signals that further attempts to activate such a link are useless, because no connection can be established at all, which differentiates it from the CNS. The generic term for CSS, CNS, and CNP is Signaling Data Link Connection Acknowledgement.

Table 8.4 (continued)

H1/H0	Abbr.	Name	Description
23	TFC	TransFer Controlled	The TFC message is sent by STPs when overload to particular Point Codes occurs. When an STP receives MSUs destined for those point codes, then the STP sends a TFC message to the sender of this MSU as a notification of the overload situation. If further MSUs are received from the same sender, then the TFC message is repeated every eighth time a MSU is received (refer also to the RCT message).
16	LIN	Link INhibit Signal	It is possible that a particular link fails frequently and, hence, must be regarded as insecure. SS7 allows to automatically take such a signaling link out of service. This is done with a LIN message. Please note that this blocks the connection for user signaling only, and that it is still possible to transfer management messages or test messages.
26	LUN	Link UNinhibit Signal	The LUN message is the counter part of the LIN message. A connection that was taken out of service was tested in the meantime, manually or automatically, and shall now be reactivated.
36	LIA	Link Inhibit Acknowledgement Signal	The LIA message is the response when a LIN message was received. A link shall not be available for user signaling anymore.
46	LUA	Link Uninhibit Acknowledgement Signal	The LUA message is the response when a LUN message was received. A connection that was taken out of service is now available again for user signaling.
56	LID	Link Inhibit Denied Signal	Negative response to a LIN message. The connection can not be taken out of service (e.g. because no alternative links are available).
66	LFU	Link Forced Uninhibit Signal	When taking a link out of service, one has to distinguish between the active side, which wants to trigger this action by sending a LIN message, and the side which receives the LIN message. Only the receiving side can send a LFU message, in order to request that this link is brought back to service. (This is a request to send a LUN message.)
76	LLT	Link Local Inhibit Test Signal	The status of the connection is tested by that SP/STP, which originally requested to inhibit the link, by periodically sending a LLT message to the peer entity. When a SP/STP receives a LLT message, it checks whether the link was actually inhibited by that node. If this is the case, no action is undertaken and the status is considered satisfactory. If, however, the link was originally inhibited by that entity which received the LLT message, then there is some inconsistency. In order to resolve this inconsistency, a LFU message is sent as an answer to the other node.

Table 8.4 (continued)

H1/H0	Abbr.	Name	Description
86	LRT	Link Remote Inhibit Test Signal	When an SP/STP receives a LIN message, it expects to periodically receive LLT messages. If this time period, during which the test message was expected expires, then a LRT message is sent to the peer SP/STP.
17	TRA	Traffic Restart Allowed Signal	The TRA message is sent by an SP/STP to a neighboring SP/STP when this neighbor comes back to service, that is, when it can carry data again. When a STP comes to service, this STP also sends TRA messages to all neighbors. This does not apply to an SP.
13	RCT	Signaling Route Set Congestion Test	The RCT message allows an SP/STP to test whether the path to a particular destination STP is available. An RCT message is sent when a TFC message was previously received from the destination STP and the state is uncertain.
1A	UPU	User Part Unavailable	When an SP receives a message for a particular User Part (e.g., SCCP) which is currently not available, then a UPU can be returned to the sender of the original message as a notification about this unavailability of the User Part.

Table 8.5
Message Types in SS7 Network Test

H1/H0	Abbr.	Name	Description
11	SLTM	Signaling Link Test Message	Layer 3 uses the SLTM for the initial test and the periodic test of a SS7 link. The SLTM carries, for this purpose, a sequence of up to 15 test bytes, which the peer needs to correctly receive.
21	SLTA	Signaling Link Test Acknowledge	Every SLTM has to be acknowledged with a SLTA. The SLTA returns the test sequence, previously received in the SLTM, back to the sender.

9

Signaling Connection Control Part

The SCCP uses the layers MTP 1 through 3 of SS7. In GSM, those services are used by a number of subsystems (Figure 9.1). The services of the SCCP are used, in particular, by the base station subsystem application part (BSSAP) on the A-interface and by the transaction capabilities application part (TCAP) together with the mobile application part (MAP) on the various interfaces within the NSS. Note that ISUP also may use the SCCP, but fewer and fewer applications use that combination.

9.1 Tasks of the SCCP

In contrast to the MTP 1 through 3, which is responsible for the transport and address functionality between two network nodes, the SCCP, by means of its Layer 3 functions, offers end-to-end addressing, even across several network nodes and countries. Additionally, the SCCP allows for a distinction among the various applications within a network node; internally, the SCCP refers to these applications as subsystems. Two connection-oriented and two connectionless service classes are available to the users of the SCCP for actual data transfer.

Furthermore, the SCCP comes with its own management functions for administrative tasks, which are independent from those known from the SS7 signaling network management. Although the SCCP is considered a Layer 3 functionality in the ITU Recommendations Q.711 through Q.714, it also provides features that belong to Layer 4, including mechanisms for error detection and an optional segmentation of the data to be transmitted.

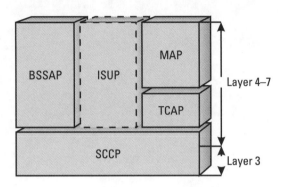

Figure 9.1 The SCCP as a platform for various users.

9.1.1 Services of the SCCP: Connection-Oriented Versus Connectionless

The SCCP offers two connection-oriented and two connectionless service classes to its users. The difference between the two is as follows. Two network nodes establish a virtual connection between the two subsystems for transaction 1, 2, or 3, in case of the connection-oriented mode. The identification of the connection is achieved via reference numbers, the source local reference (SLR) and the destination local reference (DLR). While such a connection is active, it is possible not only to exchange data between the two network nodes but also to address individual transactions.

Figure 9.2 illustrates this relation. The SCCP analyzes the data received from the MTP and forwards the data to the addressed subsystem, where the input data is associated with the various active transactions. Typical examples in GSM for connection-oriented transactions are a location update and a MOC within the BSSAP.

In the case of connectionless service classes, the SCCP provides no referencing; the recipient of a message must assign it to an active process. Examples for connectionless applications are PAGING in the BSSAP, SCCP management, and the TCAP protocol.

Distinguishing between connection-oriented and connectionless service within the SCCP is achieved by a parameter called the protocol class (described in Section 9.3.2.5).

9.1.2 Connection-Oriented Versus Connectionless Service

The difference between connection-oriented and connectionless service can best be explained by the example of sending a letter. The postal service provides the physical means for mail transfer. The individual envelopes correspond to

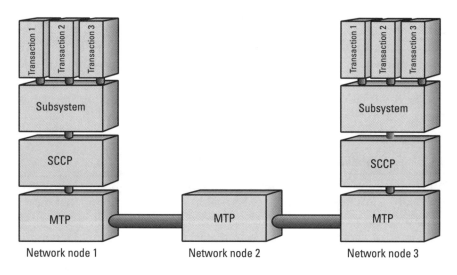

Figure 9.2 Connection-oriented services of the SCCP.

the MSUs, and the letter inside the envelope corresponds to the SCCP message (Figure 9.3).

9.1.2.1 Connection-Oriented Service

When two parties of any particular company correspond via mail, they typically address many issues. References for each issue need to be assigned, so the recipient can distinguish among them. That corresponds to a virtual

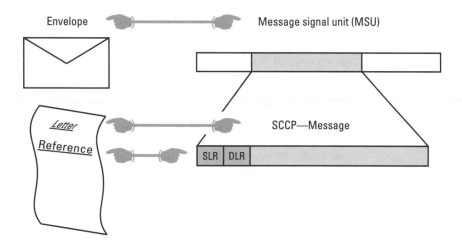

Figure 9.3 The task of SLR and DLR.

connection setup. The various issues that arise could be an unpaid bill or a new order. Each side tries to make the issue clear, for example, by adding a headline or a reference line to establish a unique reference. The function of the reference corresponds to the task of SLR and DLR of connection-oriented services in the SCCP. A virtual connection between sender and recipient is set up in both cases. "Virtual" here means that no permanent, dedicated physical path between the two parties exists.

9.1.2.2 Connectionless Service

A person who vacations in a faraway country typically sends postcards to relatives and friends. Each postcard needs an address to enable delivery, but there is no reference to a specific issue and no answer is expected. It is, therefore, a conversation that does not require an immediate reference or a connection setup.

This comparison is valid only for the SCCP itself. The recipient has the opportunity to include a reference in the data part of the message and hence establish a relation to an issue, even when using the connectionless service classes. (An example is provided in Chapter 11.)

9.2 The SCCP Message Format

The complete SCCP message is hosted, together with the routing label by the SIF of an MSU (Figure 9.4). Only the identifier for the user part SCCP is carried in the SIO outside the SIF. The SCCP is the immediate layer above the MTP, and a wide variety of messages with different formats and tasks are defined for the SCCP. The peculiarities of the SCCP message format will be explained first; the single messages are then described in more detail. Figure 9.5 presents the general format of a SCCP message.

SCCP messages consist of the following parts (refer to Figures 9.4 and 9.5):

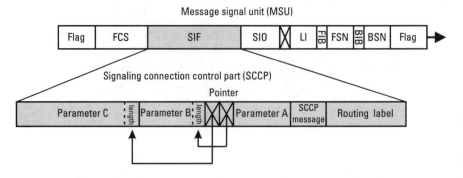

Figure 9.4 The MSU as the transport frame for the SCCP.

Signaling Connection Control Part

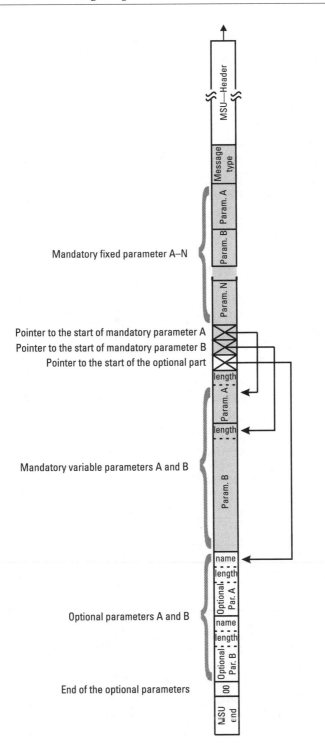

Figure 9.5 General format of an SCCP message.

- Mandatory fixed part. The parameters of this part are mandatory and of fixed length, and their order is fixed. That allows omission of an identifier for the parameter as well as a length indicator.
- Mandatory variable part. The parameters of this part are mandatory and their order is fixed; however, their length may vary, depending on the situation. Again, no identifier is necessary, but a length indicator is required to determine the parameter's position in the message. The length indicator uses an additional byte for each parameter.
- Optional part. All the parameters of this part are optional, that is, a particular parameter may or may not be present in a given message, depending on the circumstances. To enable the recipient of a message to identify the optional parameters, they require an identifier and a length indicator for each such parameter present in the message. Every SCCP message that can contain optional parameters has to have an end-of-optional-parameters (EO) indicator to signal the end of the parameter list (see also Section 9.3.2.3). The code for the EO is 00, which mandates that this value be excluded as a valid identifier.
- Pointer. Every pointer is 1 byte in length. The value of a pointer indicates the distance to the beginning of the field to which it points. One pointer is necessary for every mandatory variable parameter, while only one pointer is necessary for the whole optional part, which points to the start of the optional part, indifferently from the number of parameters contained in that part.

Figure 9.5 presents the general format of a SCCP message. The mandatory part is shaded, while the optional part is in white. (Similar shading is applied to all illustrations of SCCP messages.)

9.3 The SCCP Messages

Figures 9.6(a) and 9.6(b) illustrate those SCCP messages used in GSM.

9.3.1 Tasks of the SCCP Messages

Table 9.1 lists all the SCCP message types that are defined in ITU Recommendations Q.712 and Q.713 *and* used in GSM. The uppercase letters relate to the abbreviations used in this context. Table 9.2 lists the SCCP management messages sent in the data part of UDT messages.

Signaling Connection Control Part

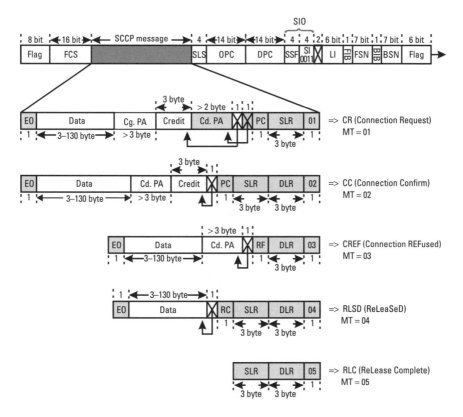

Figure 9.6(a) SCCP messages in GSM (part 1).

9.3.2 Parameters of SCCP Messages

This section presents all the parameters of SCCP messages and describes their task. The same abbreviations are used in the description as in Figure 9.6(a) and 9.6(b). When length information is given for a parameter, it relates only to the parameter itself and not to a possibly necessary length indicator or type field (different from the illustrations).

9.3.2.1 Calling-Party Address and Called-Party Address (>2 Bytes)

The calling-party address (CaPA) and the called-party address (CdPA) have the same format and identify the type of address as well as the address itself. An address may consist of any combination of the following:

- The SPC (2 bytes);
- The SSN (1 byte);
- The global title (> 3 bytes).

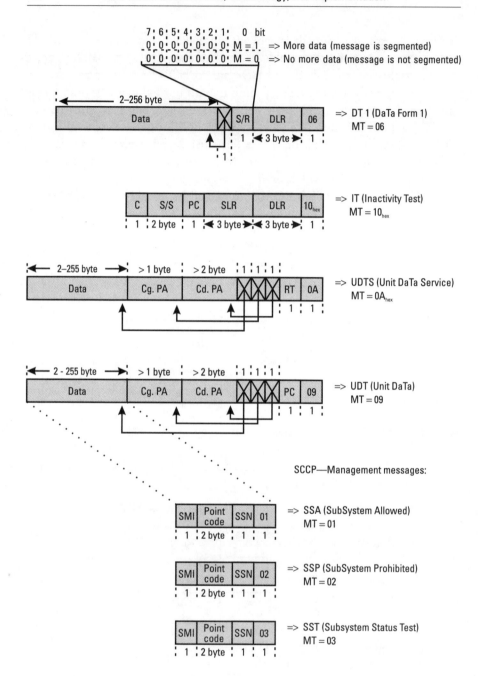

Figure 9.6(b) SCCP messages in GSM (part 2).

Table 9.1
SCCP Message Types

ID (Hex)	Message Type	Connection Oriented?	Description
01	CR (Connection Request)	Yes	Is sent from the BSC to the MSC or vice versa at the beginning of a connection set up, in order to request an SCCP connection. A CR includes in its data part, for example, the whole LOC_UPD_REQ or HND_REQ (BSSAP).
02	CC (Connection Confirm)	Yes	A positive response to CR. Acknowledges receipt of CR and establishment of the requested SCCP connection.
03	CREF (Connection REFused)	Yes	Negative response to CR. The SCCP of a signaling point (BSC or MSC) is unable to provide the requested SCCP connection. A cause value is supplied when the CR is answered by CREF.
04	RLSD (ReLeaSeD)	Yes	The RLSD message is always sent from the MSC to the BSC, in order to release an SCCP connection. The assigned memory resources are released, too.
05	RLC (ReLease Complete)	Yes	The acknowledgement of the receipt of an RLSD message and the confirmation that the assigned SCCP resources were released.
06	DT 1 (DaTa Form 1)	Yes	The entire data transfer between BSC and MSC is performed in DT 1 messages after an SCCP connection was established by a CR and CC. DT 1 messages belong, in contrast to DT 2 messages (which are not used in GSM), to protocol class 2. DT 2 messages belong to protocol class 3 and provide additional mechanisms for error detection.
09	UDT (Unit DaTa)	No	In contrast to the messages presented above, UDT messages provide for connectionless services (protocol class 0 and 1). UDT messages in GSM are used by MAP/TCAP for all data transfer tasks and on the A-interface to convey PAGING messages (among others). Another application of UDT messages is to transmit SCCP management messages.
0A	UDTS (Unit DaTa Service)	No	When the SCCP of a signaling point receives a UDT message with protocol class 1 that can not be processed or forwarded to the addressed subsystem, then a UDTS message is returned to the sender. The original UDT message may then be repeated.

Table 9.1 (continued)

ID (Hex)	Message Type	Connection Oriented?	Description
10	IT (Inactivity Test)	Yes	Every side may periodically send an IT message in order to query the state of an SCCP connection and correct a possible inconsistency of data.

Table 9.2
SCCP Management Messages
(Sent in the Data Part of UDT Messages)

ID (Hex)	Message Type	Connection Oriented?	Description
01	SSA (SubSystem Allowed)	No	Both sides may send it to the respective peer as information that a previously not available subsystem is now available. Examples of subsystems are SCCP management, BSSAP, the VLR, the HLR, or the MSC.
02	SSP (SubSystem Prohibited)	No	An available subsystem has to be taken out of service.
03	SST (SubSystem Status Test)	No	The state of a subsystem which is reported as not available can be queried by sending an SST message.

If all three are present, they appear in exactly the order listed. Figure 9.11 presents the subsystems currently defined for the SCCP. The parameters CaPA and CdPA are necessary for end-to-end addressing of SCCP messages, as indicated in Figure 9.7.

MAP uses all possible combinations, while the BSSAP requires only the SPC and the SSN (= BSSAP) for addressing.

Global Title

A switching exchange or any SCCP network node, particularly for international connection requests, has no information source for routing purposes

Signaling Connection Control Part

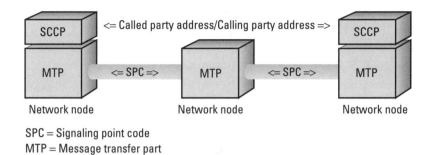

SPC = Signaling point code
MTP = Message transfer part

Figure 9.7 End-to-end addressing of the SCCP.

other than the dialed directory number. That is exactly what a global title is. The global title consists of a regular directory number and information as to how to interpret that number. The example in Figure 9.8 shows a called-party address with a global title taken from a MAP trace. Not a subscriber but a particular VLR of the D1 network in Germany is addressed in the figure. The addressing scheme is formatted according to ITU Recommendations E.163 and E.164. The ISDN address of the VLR is 49 171 062 6666.

```
Called Party Address
    reserved for national use : 0
    routing indicator : routing based on global title
    global title indicator : 4 = global title includes translation
    type,numbering plan,encoding scheme and nature of address indicator
    SSN indicator : address contains a subsystem number
    point code indicator : address contains no signaling point code
    subsystem number : 7 = GSM-VLR
    translation type : 0
    numbering plan : 1 = ISDN/telephony numbering plan (recommendation
    E.163 and E.164)
    encoding scheme : 2 = BCD, even number of digits
    nature of address indicator : 4 = international number
    address information : 491710626666
```

Figure 9.8 Example for a CdPA with global title.

Format of Calling-Party Address and Called-Party Address

Figure 9.9 presents the format of CaPA and CdPA as used by the MAP. This illustration also is valid for other applications.

- The first byte defines the structure of the remainder of the information, that is, how the following data shall be processed. Bits 0 through 5 of the first byte indicate whether a parameter is included and what parts the global title consists of, if included. Bit 6 of the first byte determines how the routing of a SCCP message shall be done. When bit 6 equals 0, the global title shall be used for routing. If bit 6 equals 1, the SSN and the SPC shall be used for routing.
- The second and third bytes carry the SPC. The two most significant bits of the SPC are coded with a 0 value, since the SPC, as defined by ITU-SS7, has only 14 bits.

7	6	5	4	3	2	1	0	bit
Reserved for national use	Routing indicator	\multicolumn{4}{c}{Global title indicator (value for MAP inter-PLMN = 0100)}			SSN indicator	Point code indicator	byte 1
\multicolumn{8}{c}{SPC (signaling point code)/bits 1–8 of 14}								byte 2
0	0	\multicolumn{6}{c}{SPC (signaling point code)/bits 9–14 of 14}							byte 3
\multicolumn{8}{c}{(SSN) subsystem number (e.g., BSSAP, EIR, HLR, SCCP management)}								byte 4
\multicolumn{8}{c}{How to translate the global title (translation type)}								byte 5
\multicolumn{4}{c}{Numbering plan (e.g. ISDN; E.163/E.164)}	\multicolumn{4}{c}{How is the global title coded (e.g., BCD, even/odd number of digits)}							byte 6
Even/odd number of digits	\multicolumn{7}{c}{Structure of the address information (international number, subscriber number, ...)}								byte 7
\multicolumn{4}{c}{2nd digit}	\multicolumn{4}{c}{1st digit}								byte 8
\multicolumn{4}{c}{4th digit}	\multicolumn{4}{c}{3rd digit}								byte 9
\multicolumn{4}{c}{......}	\multicolumn{4}{c}{.........}								...
\multicolumn{4}{c}{'1111' (fill digit ↔ in case of odd number of digits)}	\multicolumn{4}{c}{last digit}								byte N

Figure 9.9 Format of a CaPA and CdPA in the SCCP.

- The fourth byte carries the SSN.
- All the remaining data belong to the global title, when present.

9.3.2.2 Credit (1 Byte)

The credit (C) parameter indicates for a secured SCCP transmission the number of SCCP messages that may be unacknowledged at any given time. Only the protocol classes 1 and 3 allow for secured transmission.

9.3.2.3 End of Optional Parameters (1 Byte)

The end-of-optional-parameters (EO) parameter is found only in SCCP messages that may contain optional parameters. It indicates the end of the part, which hosts all the optional parameters of a SCCP message and, hence, the end of the SCCP message as such.

9.3.2.4 Message Type (1 Byte)

The message type (MT) parameter defines the SCCP message type. Its values are presented in Figures 9.6(a) and Figure 9.6(b).

9.3.2.5 Protocol Class (1 Byte)

The protocol class (PC) parameter indicates the service class of a message. Four protocol classes are defined (0, 1, 2, 3), where 0 and 1 represent the connectionless services, while 2 and 3 represent the connection-oriented services. Not all messages can be sent in any service class.

Two protocol classes are defined for both connection-oriented and connectionless service, where classes 0 and 2 form the basic version and classes 1 and 3 allow for additional data security in the form of acknowledgements. Over the A-interface, only protocol classes 0 and 2 are used, while the GSM MAP uses protocol classes 0 and 1. The protocol class 3 is not used at all by GSM.

It is important to distinguish between connection-oriented and connectionless service classes, when analyzing the PC parameter (Figure 9.10). Bits 0 through 3 describe, in both cases, the actual protocol class. Bits 4 through 7 are

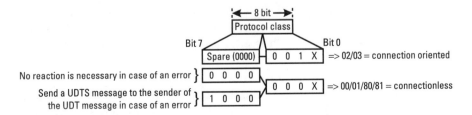

Figure 9.10 Possible formats of the SCCP PC parameter.

not used, in the case of connection-oriented service classes 2 and 3, while bits 4 through 7, in the case of connectionless service classes 0 and 1, indicate whether a UDT message has to be answered in case of an error.

9.3.2.6 Release Cause (1 Byte)

The release cause (RC) parameter provides information about the cause of a SCCP connection being released. Note that the value $0F_{hex}$ (unqualified) is used in the case of normal clearing of a SCCP connection.

9.3.2.7 Refusal Cause (1 Byte)

The refusal cause (RF) parameter provides the reason for a request for a SCCP connection setup being refused. There is, in particular, a distinction between overload of the SCCP and overload of a subsystem.

9.3.2.8 Return Cause (1 Byte)

The return cause (RT) parameter is present only in UDTS messages. It indicates why the sender of a UDTS message could not process a UDT message.

9.3.2.9 Source Local Reference/Destination Local Reference (Each 3 Bytes)

The source local reference (SLR) and destination local reference (DLR) parameters are present only in connection-oriented messages. Their task is to identify a virtual connection.

9.3.2.10 Segmenting/Reassembling (1 Byte)

The segmenting/reassembling (S/R) parameter is present only in DT 1 messages. The data part of DT 1 messages is limited to 256 bytes. When information larger than 256 bytes has to be transmitted, the data has to be segmented, that is, distributed over several DT 1 messages (compare the segmentation of $LAPD_m$, described in Chapter 7).

9.3.2.11 Sequencing/Segmenting (2 Bytes)

The sequencing/segmenting (S/S) parameter is used only in the DT 2 message and the IT message. It contains SCCP internal information, the send sequence number P(S) and the receive sequence number P(R), as well as information on segmentation, which is similar to S/R. The task of P(S) and P(R) corresponds to that of N(S) and N(R) of LAPD, or FSN and BSN in SS7 respectively.

9.3.2.12 Subsystem Number (1 Byte)

The subsystem number (SSN) specifies the user from which a SCCP message stems or to which it is addressed. The SSN may be part of CaPA/CdPA or part of an SCCP management message. Table 9.3 lists the subsystems of the SCCP.

Table 9.3
Subsystems of the SCCP

SSN (Hex)	Subsystem
00	SSN not known or not available
01	SCCP management
02	Reserved
03	ISUP
04	OMAP
05	MAP
06	HLR
07	VLR
08	MSC
09	EIR
0A	AuC (future)
FE	BSSAP

9.3.3 Decoding a SCCP Message

Figure 9.11 shows a RLSD message that was recorded by a protocol tester. The explanation is intended to clarify the relationship between the hexadecimal and the mnemonic representations.

9.4 The Principle of a SCCP Connection

The same principle is always used to establish and release an SCCP connection. Figure 9.12 illustrates that principle for the BSC-to-MSC example. One side requests a SCCP connection by sending a CR message. An important part of the CR message is the SLR, which is used as a reference by the requesting side like some kind of ticket number, to identify the requested SCCP connection.

The peer entity confirms the establishment of the SCCP connection by sending a CC message. The CC message contains the SLR of the responding side and the DLR (which corresponds to the original SLR of the requesting side). Hence, the SLR on the side of the BSC corresponds to the DLR on the MSC side and vice versa. Both sides know the SLR, DLR, and when the CC message was received, and the two parameters are then used as identifiers for sender and addressee.

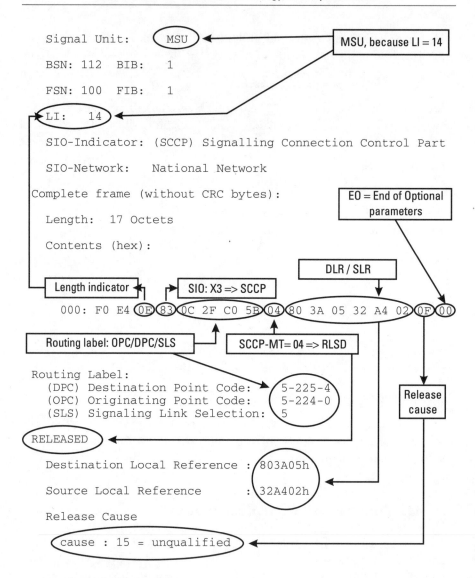

Figure 9.11 An RLSD message.

When a connection is set up and each side knows the "ticket number" (the reference number (DLR) of the peer entity) the DT 1 message is used for further exchange of data (the DT 2 message is not used by GSM).

Note that the establishment of a radio connection or a call on the A-interface is performed via DT 1 messages. This shows the transparency of

Signaling Connection Control Part

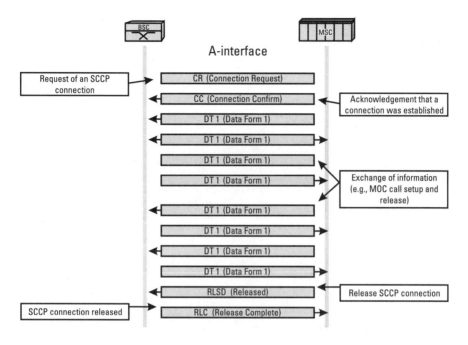

Figure 9.12 Principle to establish and release an SCCP connection for BSC-to-MSC.

the SCCP for its users (the BSSAP is this user on the A-interface). Only after release of the radio connection is the SCCP connection also released.

An SCCP connection is released by an RLSD message, which in the BSC-to-MSC example is always sent by the MSC. The receiving side confirms with an RLC message that the RLSD message was received, the SCCP connection was released, and the resources were cleared.

10

The A-Interface

On the physical level, the A-interface consists of one or more PCM links between the MSC and the BSC, each with a transmission capacity of 2 Mbps. The TRAU, which typically is located between the MSC and the BSC, has to be taken into consideration when examining this interface. Consequently, the A-interface can be separated into two parts.

- The first part is between the BTS and the TRAU, where the transmitted payload still is compressed. Figure 10.1 shows a possible channel configuration for three trunks. As on the Abis-interface, a single traffic channel occupies only two of the eight bits of a PCM channel. That is why it is possible to transport four fullrate traffic channels on one PCM channel. The exceptions are TSs, where signaling information is carried. Signaling information requires the entire 64 Kbps of a channel (e.g., TS 16 in Figure 10.1).
- The second part of the A-interface is the one between the TRAU and the MSC. There, all data are uncompressed. For that reason, every traffic channel requires the complete 8 bits or occupies an entire 64-Kbps PCM channel. The position of a signaling channel may be different before and after the TRAU, as Figure 10.1 shows.

10.1 Dimensioning

An examination of the channel configuration in Figure 10.1 makes it obvious that the transmission resources between the BSC and the TRAU are used only

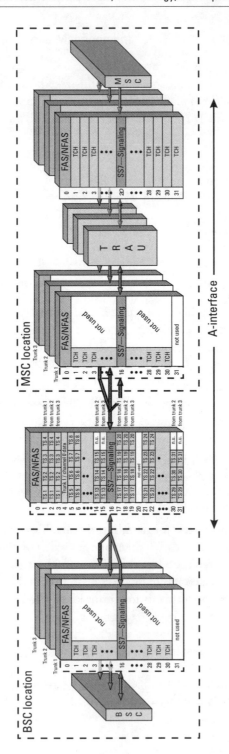

Figure 10.1 Possible channel configuration between the BSC and the MSC.

inefficiently. Only 2 bits of the PCM channel are actually occupied, while the remaining 6 bits stay vacant. In that respect, the A-interface between the BSC and the TRAU can be compared to the Abis-interface in a star configuration.

It is possible, by means of multiplexing, to transport the data of several trunks from the BSC over only one physical 2-Mbps link before the data are actually handed over to the TRAU for decompression. That allows a savings of about two-thirds of the line costs if the TRAU is installed at the MSC location. Multiplexing between the TRAU and the MSC does not deserve any consideration, since every traffic channel there requires 64 Kbps.

10.2 Signaling Over the A-Interface

As on all the other interfaces except for the Air-interface and the Abis-interface, the A-interface uses SS7 with the SCCP as the user part. Similar to the Abis-interface, GSM uses an already existing signaling standard (SS7 plus SCCP) on the A-interface and simply defines a new application.

This new application is the BSSAP, which itself can be separated into the base station subsystem management application part (BSSMAP) and the direct transfer application part (DTAP).

That results in the OSI protocol stack are presented in Figure 10.2. DTAP data are user information and therefore completely transparent for the A-interface, while the BSSMAP data are part of Layer 3.

10.2.1 The Base Station Subsystem Application Part

The BSSAP, with its two parts, the DTAP and the BSSMAP, represents the GSM-specific user signaling on the A-interface.

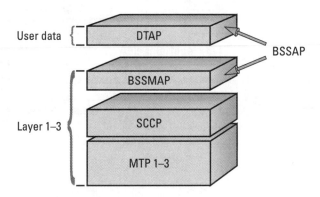

Figure 10.2. The A-interface in the OSI protocol stack.

- The BSSMAP includes all messages exchanged between the BSC and the MSC that are actually processed by the BSC. Examples are PAGING, HND_CMD, and the RESET message. More generally, the BSSMAP comprises all messages exchanged as RR messages between the MSC and the BSC, as well as messages used for control tasks between the BSC and the MSC.
- The DTAP comprises all messages exchanged between a subsystem of the NSS and the MS. The messages are transparent for the BSS. This definition applies to all but three messages of MM. These exceptions are the LOC_UPD_REQ, the IMSI_DET_IND, and the CM_SERV_REQ messages. All three are part of DTAP but nevertheless are partly processed by the BSC.

Figure 10.3 illustrates the task sharing between BSSMAP and DTAP. It is important to note that there is not a 100% correspondence between DTAP and CC/MM on one side and BSSMAP and RR on the other. There are cases when the BSC and the MS exchange RR messages without informing the MSC about the content of the messages. The same applies for BSSMAP messages between the BSC and the MSC.

10.2.2 The Message Structure of the BSSAP

Figure 10.4 shows the general structure of BSSAP messages. The entire BSSAP message is embedded in an SCCP message. The first 8 or 16 bits of the BSSAP distinguish between BSSMAP and DTAP. The DTAP header is

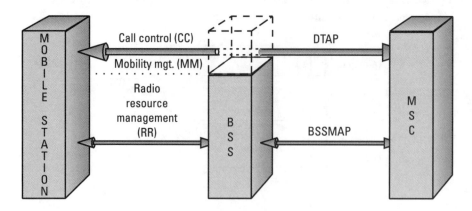

Figure 10.3 The relation between DTAP corresponding to CC and MM, and BSSMAP corresponding to RR.

The A-Interface 175

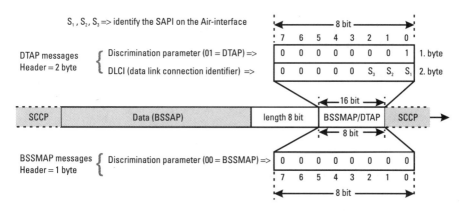

Figure 10.4 The format of BSSAP messages.

2 bytes (16 bits) long and consists of the discrimination parameter (01 = DTAP) and the data link connection identifier (DLCI). The first 3 bits of the DLCI identify the service access point identifier (SAPI), which is used on the Air-interface (SAPI = 0 for RR, MM, and CC; SAPI = 3 for SMS and SS).

The BSSMAP header is only 1 byte (8 bits) long and consists only of the discrimination parameter (00 = BSSMAP). In BSSMAP, there is no DLCI octet.

A length indicator, indicating the length of the following data field, follows the header in both cases of BSSMAP and DTAP. DTAP messages exactly follow the format as presented in Figure 7.14. BSSMAP messages are formatted as shown in Figure 10.5. The actual parameters follow the 1-octet-long message type. Independent of being mandatory or optional, each parameter always consists of an information element identifier (IEI), length indicator, and the actual data.

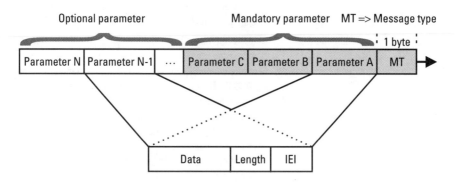

Figure 10.5 The internal structure of BSSMAP messages.

10.2.3 Message Types of the Base Station Subsystem Management Application Part

Table 10.1 lists all BSSMAP messages, along with brief descriptions of their tasks. The uppercase letters indicate the abbreviations used in this context.

Table 10.1
BSSMAP Messages

ID (Hex)	Name	Direction	Description
01	ASSignment REQuest	MSC → BSC	Is sent from the MSC to the BSC during a connection set up in order to assign a channel on the Air- and the A-interface. The ASS_REQ message does not specify the Air-interface channel in more detail. Rather, the BSC selects one TCH out of the available radio resources and assigns this channel by means of the ASS_CMD message. The ASS_REQ message, however, contains a specification of the channel on the A-interface. There is no TCH available on neither the Air-interface nor on the A-interface before the ASS_REQ message is received and the complete communication between NSS and MS occurs via control channels.
02	ASSignment COMplete	BSC → MSC	This is the positive response to an ASS_REQ for TCH assignment. The MS has changed to the TCH and the Layer 3 connection is established.
03	ASSignment FAILure	BSC → MSC	This is the negative response to the MSC, when a channel assignment was not successful (not used for errors during handover). ASS_FAIL should not be confused with ASS_FAI, a message, defined in GSM 04.08 for the Air-interface. The cause values, for example, are different between the two messages.
10	HaNDover REQuest	MSC → BSC	If the BSC needs to be changed during handover, this message is sent by the MSC to the new BSC. The MSC reacts with HND_REQ to HND_RQD originating from the old BSC. The new BSC selects the appropriate radio resources and sends the detailed information (e.g., frequency, time slot, handover reference) in a HND_REQ_ACK message, back to the MSC. The HND_REQ message, like the ASS_REQ, contains a specification of the resources to be used on the A-interface.

Table 10.1 (continued)

ID (Hex)	Name	Direction	Description
11	HaNDover ReQuireD	BSC → MSC	The BSC uses this message to request a handover from the MSC (only intra-MSC and inter-MSC handover). The best candidates for the handover are the most important content. The possible destinations are derived from the measurement results of the MS and neighbor cell relations of the serving cell.
12	HaNDover ReQuest ACKnowledge	BSC → MSC	Answer of the BSC to a HND_REQ message. The HND_REQ_ACK message contains the HND_CMD message, which was prepared by the new BSC and shall be sent to the MS via the MSC and the old BSC.
13	HaNDover CoMmanD	MSC → BSC	Is used for every handover that is controlled by the MSC, in order to provide detailed information about the new radio resources to the MS. Please note that this HND_CMD message has a different format compared to the one with the same name, defined by GSM 04.08 for the Air-interface.
14	HaNDover CoMPlete	BSC → MSC	A HND_CMP message has to be sent to the MSC for every successful handover, which is controlled by the MSC. In case of an intra-BSC handover (which most likely is controlled by the BSC), HND_PERF is used to inform the MSC about the change of channels within the BSC, instead of HND_CMP.
16	HaNDover FAILure	BSC → MSC	A HND_FAIL is always sent from the BSC to the MSC when a handover fails, e.g., due to insufficient resources (answer to HND_REQ) or when handover as such fails (answer to HND_CMD).
17	HaNDover PERFormed	BSC → MSC	Is sent to the MSC for every handover, which is controlled by the BSC (only intra-BTS and intra-BSC) in order to inform the MSC about a channel change. HND_PERF corresponds to HND_CMP in case of the MSC controlled handover.
18	HaNDover CaNDidate ENQuire	MSC → BSC	There might be cases when it is required to lower the traffic load on one or more cells. For this purpose, the MSC has the ability to send a HND_CND_ENQ message to the BSC. The message identifies the cell, where the load shall be reduced and possible neighbor cells, and where already active calls could be transferred. For every connection, which can—from a radio perspective—be transferred, the BSC responds with a HND_RQD message to the MSC.

Table 10.1 (continued)

ID (Hex)	Name	Direction	Description
19	HaNDover CaNDidate RESponse	BSC → MSC	With an HND_CND_RES message, the BSC confirms to the MSC that a HND_CND_ENQ message was received and processed. The confirmation that the message was processed refers, in particular, to the fact that HND_RQD messages were sent for all possible candidates for handover.
1A	HaNDover ReQuireD REJect	MSC → BSC	This message is only sent by the MSC as a response to HND_RQD when: a) the HND_RQD was not processed within the time defined by timer T7 (i.e., if no HND_CMD is sent) and b) the Response Request parameter was set in the HND_RQD message. In this case, the MSC has to answer each not processed HND_RQD. If the Response Request parameter is not active then the BSC may simply repeat the HND_RQD after the expiration of timer T7.
1B	HaNDover DETect	BSC → MSC	The BSC reacts with this message when it receives a HND_DET from the BTS (same name as the message on the Abis-Interface). The HND_DET message informs the MSC at the earliest possible time about a change of the radio resources. This measure, on the other hand, allows for rapid switch of the terrestrial resources (reduction of "dead air"-time during handover).
20	CLeaR CoManD	MSC → BSC	This message is always used to release the radio resources to a specific MS. A CLR_CMD may be: a) the answer to a CLR_REQ, that is, a problem that was detected by the BSS; (b) the reaction to a problem detected by the NSS; or (c) used in case of normal termination of the radio resources. Only further analysis provides details on the reason for a CLR_CMD.

Table 10.1 (continued)

ID (Hex)	Name	Direction	Description
21	CLeaR CoMPlete	BSC → MSC	When the BSC receives a CLR_CMD message it clears the radio resources to a particular MS. This is confirmed by sending a CLR_CMP message. Sometimes, the CLR_CMP message is sent even before the radio resources are actually released, i.e., before the RF_CHAN_REL message is sent on the Abis-interface.
22	CLeaR REQuest	BSC → MSC	A CLR_REQ message with the appropriate cause is sent to the MSC, if the BSC detects severe problems with an existing connection to a MS on a control channel or a TCH. The BSC reacts with a CLR_REQ message when a CONN_FAIL message is received from the BTS, but also in case of some error situations during handover. The MSC reacts with a CLR_CMD message (for more details see Chapter 13, "Quality of Service").
25	SAPI"n" REJect	BSC → MSC	The BSC responds with a SAPI_REJ message if it receives a DTAP message from the MSC with a SAPI value other than zero, and no corresponding connection to a MS exists or can be established. This message contains an appropriate cause value and the 'wrong' DLCI (Data Link Connection Identifier).
26	CONFUSION	BSC ↔ MSC	When the BSC or the MSC receives a message with apparently the wrong content in the BSSAP header (protocol error), then a CONFUSION message is sent back to the sender. This message contains a diagnosis element that allows the sender of the faulty message to draw conclusions on the type of problem.
30	RESET	BSC ↔ MSC	In case of fatal errors, which reveal serious inconsistencies regarding the communications data between BSC and MSC (used SCCP reference, data about active calls, etc.), the reset-procedure is performed. The RESET message is then used to synchronize the BSC and the MSC. It is sent in connectionless mode (protocol class 0) in a UDT-SCCP message by that network element that detects the inconsistency. The RESET message is also utilized when the A-interface is originally initialized in order to bring both sides into a defined state. (Note that after the reset-procedure, the BSC has to repeat BLO/CIR_GRP_BLO messages, for all channels, which were in a blocked state before a RESET was sent.)

Table 10.1 (continued)

ID (Hex)	Name	Direction	Description
31	RESet ACKnowledge	BSC ↔ MSC	The RES_ACK message confirms that a RESET message was received and that all resources were actually reset.
32	OVERLOAD	BSC ↔ MSC	The BSC sends this message to the MSC in order to indicate an overload situation in a BTS or even the whole BSS. It is possible to specify the type of overload and which BTSs are affected. This message can be sent by the MSC as well, e.g., in order to indicate processor overload. The OVERLOAD message is sent to the peer within a UDT-SCCP message (connectionless mode, protocol class 0).
34	RESet CIRCuit	BSC ↔ MSC	The RES_CIRC message is used like the RESET message, to reset resources between BSC and MSC. However, the RES_CIRC message applies only to individual time slots on the A-interface, while the RESET message is used on a per trunk basis. Therefore, the RES_CIRC message contains a parameter that identifies the respective time slot.
35	RESet CIRCuit ACKnowledge	BSC ↔ MSC	The RES_CIRC_ACK message confirms to the peer entity that a RES_CIRC message was received and the respective channel was reset.
36	MSC INVoke TRaCe	MSC → BSC	GSM allows to track a single connection from the beginning to the end. For this purpose, the OMC allows the activation of a supervisory function, which translates on the message level in the direction from MSC to BSC into an MSC_INV_TRC message (MSC BSC) and in the reverse direction, from the BSC to the MSC into a BSS_INV_TRC message (BSC MSC). The connection to be supervised is determined by SLR/DLR, with which the message is sent. Other parameters, like the type of connection, identity of the MS, and identity of the OMC are included.
37	BSS INVoke TRaCe	BSC → MSC	See MSC_INV_TRC.
40	BLOck	BSC → MSC	Individual traffic channels sometimes need to be disabled for traffic. Like in ISUP, this request is sent in a BLO message, which the BSC sends to the MSC. The BLO message allows to single out an individual channel.
41	BLOcking ACKnowledge	MSC → BSC	This is the acknowledgment that a BLO message was received and processed. The channel indicated in the BLO message is no longer assigned.

Table 10.1 (continued)

ID (Hex)	Name	Direction	Description
42	UnBLOck	BSC → MSC	The UBLO message is used to cancel the blockage of a single channel on the A-interface. Hence, the UBLO message is the counter part of the BLO message.
43	UnBLOcking ACKnowledge	MSC → BSC	This is the acknowledgment that a UBLO message was received and processed. The channel, indicated in the UBLO_ACK message, can now be assigned again.
44	CIRcuit GRouP BLOck	BSC → MSC	Frequently, not only a single channel needs to be blocked, but a complete PCM trunk. If a single BLO message had to be sent for each individual time slot of a trunk, then the SS7 system would experience unnecessary load, or even overload. Therefore, the CIRC_GRP_BLO message allows to block a complete area or a whole PCM link.
45	CIRcuit GRouP BLOcking ACKnowledge	MSC → BSC	Acknowledgement by the MSC that it received and processed a CIRC_GRP_BLO message. The area or the trunks, which were specified in the CIRC_GRP_BLO message will no longer be assigned.
46	CIRcuit GRouP UnBLOck	BSC → MSC	The CIRC_GRP_UBLO message is used to cancel the blockage of an area on the A-interface. The CIRC_GRP_UBLO message is the counterpart to the CIRC_GRP_BLO message.
47	CIRcuit GRouP UnBLOcking ACKnowledge	MSC → BSC	Acknowledgement by the MSC that a CIRC_GRP_UBLO message was received and processed. The area, defined in the CIRC_GRP_UBLO_ACK message can now be assigned again.
48	UNEQipped CIRcuit	BSC ↔ MSC	If the BSC or the MSC receives a message, e.g., RES_CIRC from its peer, where channels are referenced that are unknown to the receiving side, then a UNEQ_CIRC message is returned.
50	RESource REQuest	MSC → BSC	When the MSC sends a RES_REQ message, it requests the BSC to provide updated information on the available radio resources of a BTS. The MSC selects the BTS and sends its identity (CI) within the RES_REQ message.
51	RESource INDication	BSC → MSC	Answer to a RES_REQ message. The RES_IND message contains all information about the radio resources of a BTS.

Table 10.1 (continued)

ID (Hex)	Name	Direction	Description
52	PAGING	MSC → BSC	In case of a Mobile Terminating Call (MTC), the MSC sends PAGING messages to all the BSCs of a location area. The particular location area is where the MS was last registered and is identified by the Location Area Code (LAC). Exactly one MS can be paged with a single PAGING message. The PAGING message always contains the IMSI, if assigned also the TMSI of the called MS.
53	CIPHER MODE CoManD	MSC → BSC	The MSC sends a CIPHER_MODE_CMD message to the BSC in order to start ciphering on the Air-interface. The most important information is the ciphering key Kc, which is required by the BTS to begin ciphering the encryption algorithm (A5/X), which is selected by the MSC/VLR. The CIPH_MOD_CMD message is another, different message of the Air-interface, which should not be confused with the one with a similar name of the A-interface, since the ciphering key Kc must not be sent over the Air-interface.
54	CLaSsMaRK UPDate	BSC ↔ MSC	It is possible that the technical information related to a MS, the Classmark information, changes during a call or just needs to be queried again. An example of such a situation when the characteristics of a mobile change during usage is when a class handheld is connected to a booster in a car. The MS sends, in this case, a CLS_MRK_UPD message to the MSC.
55	CIPHER MODE CoMPlete	BSC → MSC	The MS confirms that a CIPHER_MODE_CMD message was received and that encryption begins.
56	QUEUing INDication	BSC → MSC	The QUEU_IND message is used only when TCH queuing is active. If this is the case, then QUEU_IND is sent to the MSC as a response to a HND_REQ or a ASS_REQ message if no radio resources are immediately available. The corresponding connection is put in a queue until a traffic channel becomes available.

Table 10.1 (continued)

ID (Hex)	Name	Direction	Description
57	Complete Layer 3 Information	BSC → MSC	The CL3I message is used to transport all initial messages by which a connection can be established (IMSI_DET_IND, PAG_RSP, CM_SERV_REQ, LOC_UPD_REQ). Note that the SCCP message, which carries a CL3I, is not a DT 1 message, but a CR message. The CL3I message is used in particular to piggyback those DTAP messages that need to be processed by the BSC. This applies for the LOC_UPD_REQ, the IMSI_DET_IND an d the CM_SERV_REQ messages.
58	CLaSsMarK REQuest	MSC → BSC	The MSC uses this message to request the MS to transmit its technical specification (Classmark information), which has possibly changed. The response by the MS is sent in a CLS_MRK_UPD message.
59	CIPHER MODE REJect	BSC → MSC	The BSC uses this message to respond to a CIPHER_MODE_CMD message if the ciphering information, received from the MSC, can not be processed properly (e.g., when the requested ciphering algorithm is not supported).
5A	LOAD INDication	BSC ↔ MSC	The BSC sends the LOAD_IND message to the MSC, in order to be forwarded to all neighboring BSCs. This message may be used to indicate an overload situation of a BTS, and is intended to influence handover decisions in the neighboring cells. The major parameters are the identity of the affected BTS and its neighbor cells, as well as the duration, for which the overload situation is anticipated to last.

10.2.4 Decoding of a BSSMAP Message

Figure 10.6 shows an extract from a trace file captured by a protocol tester. It shows a CLR_CMD message in both hexadecimal and decoded forms.

The intention here is to emphasize the message structure of BSSMAP, where a number of parameters follow the message type identifier. These parameters are, in the case of CLR_CMD, the two information elements: Layer 3 header information and cause.

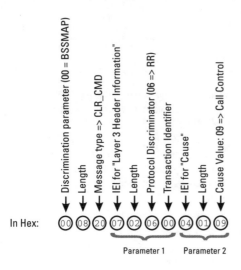

Figure 10.6 Decoding of a CLR_CMD message.

- The Layer 3 header information contains the PD and the TI that are to be used on the Air-interface.
- The cause identifies the reason why a specific radio resource will be released. Normal values are 09, which stands for CC and indicates that CC requests the release of a connection when the call is terminated, and $0B_{hex}$, which indicates a successful handover.

11

Transaction Capabilities and Mobile Application Part

The TCAP and the MAP are "on top" of SS7 (MTP 1–3) and SCCP, the basis for signaling on all NSS interfaces. Both are dealt with in this one chapter, because the functionality of the MAP cannot be understood without knowing about the TCAP. The TCAP, with its Layers 4 through 6 provides the GSM-specific MAP with a standardized interface to the transmission medium and to SS7. The clearly separated border between TCAP and MAP, as shown in Figure 11.1, is in practice more difficult to identify. The transition between the two layers is rather fuzzy. An essential precondition to understanding MAP is the study of TCAP. Above MAP, there are the applications themselves, in the GSM case there are the NSS subsystems HLR, VLR, MSC, and EIR.

11.1 Transaction Capabilities Application Part

TCAP uses SS7 or, more precisely, the SCCP, as shown in Figure 11.1. The TCAP protocol is, to some extent, the most important piece of the protocol stack for GSM or any other mobile system, because it provides the core functionality to support roaming.

Like the SCCP, TCAP is not restricted to being used by only mobile services but is utilized by many other applications for database access and similar tasks. In that respect, TCAP is different from all previously presented protocols. TCAP allows its users to access databases and switching exchanges via the worldwide SS7 network and to invoke services or modify parameters. That does

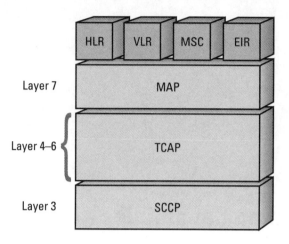

Figure 11.1 Positioning of MAP and TCAP in the SS7 protocol stack.

not exclude TCAP from being used as a platform for pure data transfer, as in GSM, after an inter-MSC handover.

TCAP is the typical implementation of the OSI layers 4 through 6. In that function, it allows integration of some translation functionality into a message, for instance, to provide a means for users of a transaction to discuss or synchronize on an application protocol. An example of this is the GSM networks of Phase 1 and Phase 2, which come with different sets of features. Therefore, those GSM networks need to exchange some information in order to synchronize the feature sets and the respective protocol elements. TCAP provides that functionality.

Figure 11.2 illustrates a generic communication process via TCAP, where, initially, both partners need to agree on the protocol to be used. The receiving side finds the respective information in the dialog control information, which in TCAP is called the dialog portion. Figure 11.2 describes the successful case only. Figure 11.2 separates the parameter part which in TCAP is called the component portion. The component portion carries the actual user data. This is MAP traffic in the case of GSM.

11.1.1 Addressing in TCAP

With respect to addressing, TCAP relies completely on the services of the SCCP. Although ITU does not explicitly exclude alternatives for the future, SCCP currently is the only platform for TCAP. TCAP uses exclusively the connectionless services of SCCP (PCs 0 and 1). The consequence is that SCCP-UDT messages are the only candidates for the transport of TCAP

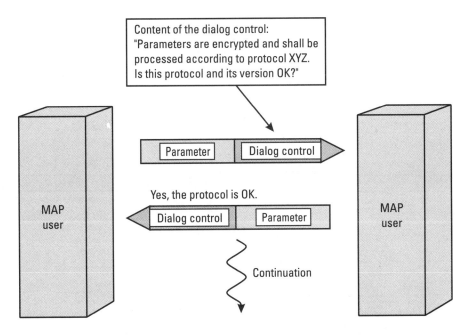

Figure 11.2 The (optional) dialog at the begin of a communication via TCAP.

messages. The sender of a TCAP message directly addresses the destination via the SCCP. The SCCP routes the message via STPs, where the actual path lies in the discretion of the SCCP.

Consider the example of addressing in TCAP/SCCP in the context of a scenario where the MSC and the HLR communicate. Figure 11.3 shows the SCCP header of a TCAP BEGIN message, where an MSC in Australia accesses an HLR in Germany. Both sender and addressee are identified by the global title. Consequently, the MSC in Australia uses the ISDN number of the HLR in Germany for addressing.

11.1.2 The Internal Structure of TCAP

TCAP can be separated into two parts or layers, as shown in Figure 11.4.

- The transaction layer in OSI Layer 4 deals with setting up and maintaining an end-to-end communication. It expects sufficient information from its user about the sender and addressee of a message. As shown in the example in Section 11.1.1, that value is not used by TCAP itself but passed to the SCCP for addressing. In most cases, the transaction layer assigns to a process an additional TCAP-internal

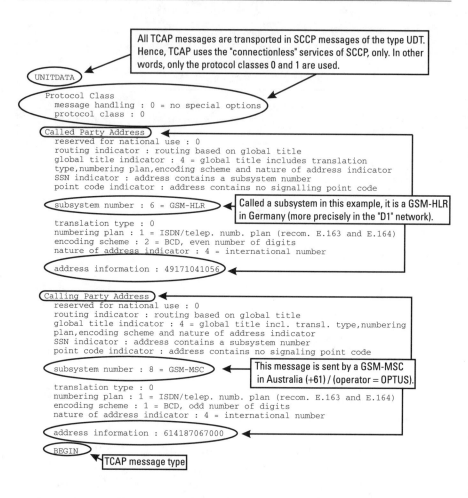

Figure 11.3 Important information in the SCCP header of a TCAP message.

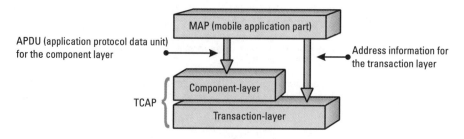

Figure 11.4 Separation of TCAP into component and transaction layers and its communication with MAP.

identifier, the transaction ID, which is comparable to SLR and DLR of the connection-oriented mode of the SCCP.
- The component layer in the OSI Layers 5 and 6 is responsible for synchronization and coordination of a communication. It also provides a uniform data interface to its users, represented by the application protocol data unit (APDU). In TCAP, APDUs are also referred to as components. They transport the payload, which MAP and the component layer exchange.

11.1.3 Coding of Parameters and Data in TCAP

One of TCAP's major advantages is its flexibility, which allows for processing of all kinds of parameter types and data formats. Take this example: TCAP (equals a shipping company) transports data (goods) of all kinds (pets, dishes, bulldozer, etc.). More technically speaking,

- TCAP has to be able to process length indicators from one byte to several thousands of bytes. That requires that a sufficiently large area is reserved for length indicators.
- It must be possible to distinguish among various parameter types. Parameter types are of little significance for the lower layers of the OSI Reference Model. In contrast, OSI Layer 6—in this case, TCAP—has the task of distinguishing and preprocessing the data for Layer 7 (MAP).

Examples for parameter types are:

- Strings (i.e., a combination of characters, e.g., "the GSM system");
- Integer numbers (0, 1, 2, 3, ...);
- Real numbers (π = 3.14159...).

Recommendations ITU X.208 and X.209 provide a complete definition of the coding of the various parameter types in ASN.1. GSM uses only a subset of those parameter types (which will be described later). There are some practical limitations with respect to the coding of parameters and length that are a consequence of the limited capacity of the data field of a UDT message (maximum 255 bytes).

In general, all data and message parts in TCAP are coded according to the same scheme (Figure 11.5), and there is no distinction between mandatory and

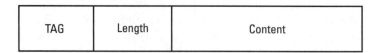

Figure 11.5 Coding of data in TCAP.

optional parameters. Every message starts with a TAG, which is an identifier, followed by a length indicator. The TAG indicates the data type of the following content.

- TAG: type, classification;
- Length: length of the content field;
- Content: The actual information.

Note that the field "content" itself also may consist of a number of TAGs, length, and content fields, which then results in an interlaced, overall structure. That can lead to a confusing structure in which significant space is consumed by type and length indicators.

Note further that the TAG field and the length indicator can be formatted in different ways, whereby the actual format is derived from the coded information and the application in use. This is more closely examined in the next section.

11.1.3.1 Formatting of the TAG Field

The TAG field is used to identify the data part, in which distinctions have to be made among data classes, formats, and types. For that reason, the TAG field itself is composed of three parts that provide the information. Note that the length indication and bit information in Figure 11.6 refer to a TAG with a length of 1 byte, only.

The meaning of the various fields is as follows:

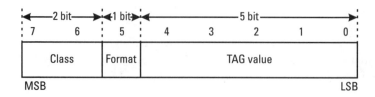

Figure 11.6 Format of TAG field (short form with length of 1 byte).

Class

Class defines the data type. Four classes need to be distinguished and are listed in Table 11.1. (The definitions provided in Table 11.1 are taken from ITU X.208.)

Format

Format distinguishes between two possible formats. It has to be noted that the distinction is valid only on the interface between MAP and TCAP.

- Format = 0_{bin}: The data field contains a primitive, which means that the parameter is not further partitioned.
- Format = 1_{bin}: The data field contains a constructor. Here, the TAG field is only a generic reference for the parameters that follow in the data field, which again are constructors or primitives.

TAG Value

The TAG value indicates to the recipient what kind of parameter type the data field carries. ITU provides a number of proposals that are mandatory within ITU applications (i.e., the universal, applicationwide, and context-specific data classes). The private-use data class can be used for proprietary data types.

11.1.3.2 Primitive Versus Constructor

The difference between a primitive, a single parameter and a constructor, and a collection of parameters is valid only in the context of formatting in TCAP. It can be explained by the example of transmitting an IMSI.

Table 11.1
Classification of Data in TCAP

Value (Bin)	Class, Explanation
00	Universal: Universal data types are specified in X.208. These data types are independent of an application, and all users of SS7 have to be able to recognize them.
01	Applicationwide: Valid only within an ITU Recommendation (e.g., TCAP message types and data types).
10	Context-specific: Valid only in an ITU application (e.g., MAP data types).
11	Private use: Network- or service-provider-specific data types, which will never be assigned by ISO or ITU.

The IMSI is a constructor per definition. It consists of MCC, MNC, and MSIN (mobile sunbscriber identification number), as presented in Figure 11.7.

In TCAP, the IMSI can be coded in two ways. Although the second way of representation may seem unusual, it still demonstrates the alternatives.

- If the IMSI is coded as a primitive, TCAP does not distinguish among MNC, MCC, and MSIN. The complete IMSI is coded as shown in Figure 11.8. The format value 0 indicates that this is a primitive and thus a single parameter message.
- If, however, the individual parameters of the IMSI are coded separately as individual parameters, then a constructor is used for the IMSI where the parameters MCC, MNC, and MSIN are the primitives (Figure 11.9). Note that the fill digit F is required for the MCC, because the MCC has a length of 3 bytes (uneven number of bytes). The following remarks are given: (1) Because the MCC, MNC, and

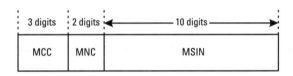

Figure 11.7 The IMSI.

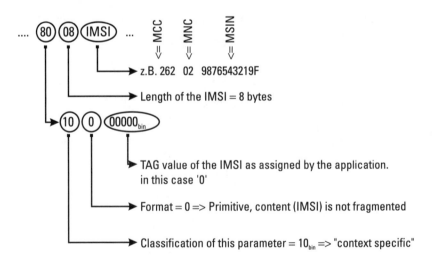

Figure 11.8 Coding of an IMSI as a primitive (with TAG and length indicator).

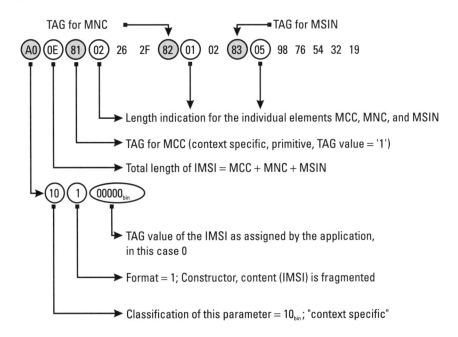

Figure 11.9 Coding of an IMSI as a constructor (with TAGs and length indicators).

MSIN are formatted as separate parameters, each requires its own TAG and length indicator, and (2) the overall length of the message increases by 6 bytes, compared to the first version.

11.1.3.3 More Options for Coding the TAG

Expansion Let us, once more, come back to Figure 11.6. The 5-bit of the TAG value field allows addressing of 31 different parameter types. That may not be enough for certain applications. Furthermore, ASN.1 has predefined most of the values (refer to the Glossary).

In addition, it may be necessary for the internal purposes of an application to assign a TAG value outside that value range (0–31).

The solution to the problem consists of the extension of the TAG to any necessary length. To do so, a special method is used, which is illustrated in Figure 11.11 and explained as follows.

- A TAG with a length of 1 byte is used for all TAG values smaller than 31_{dez}. The TAG value is binary coded. Hence, the maximum TAG value is 30_{dez} and its binary representation is 11110_{bin}.

- If the TAG value exceeds 30_{dez}, then more than one octet is required to code that value. Therefore, the value 11111_{bin} for the first byte of the TAG is reserved to indicate that the TAG is extended. In this way of coding, bits 0 through 6 of the following octets contain the actual value of the TAG. To be more precise, bits 0 through 6 represent the TAG value, while bit 7 always indicates whether another octet with a TAG value field follows. If bit 7 is set to 1, the next octet also contains TAG information; if bit 7 is set to 0, the TAG ends with this octet. Bit 6 of the *second* octet is the most significant bit (MSB), while bit 0 of the *last* octet represents the least significant bit (LSB).

Data Type Octetstring Two variants of TAG coding have been presented, the "short" and the "long" versions, which can be assigned by the user based on data type and TAG value. GSM uses yet another TAG borrowed from ASN.1. This data type is the octetstring, which is always used as the TAG when the data type does not require that an explicit identification be provided.

Data type octetstring has a fixed TAG value of 00100_{bin}, which is the representation of 4_{dez}.

For the class = universal = 00 and format = primitive = 0, the result for the TAG is 04_{hex}.

The data type octetstring was defined by ITU, in particular, to transport strings, where the individual characters are ASCII coded. The Glossary provides a complete list of all variable types and the assigned TAGs.

An example of "GSM" coded as octetstring in shown in Figure 11.10.

Note a peculiarity of MAP when it uses the octetstring TAG. When numbers need to be transmitted, the respective digits are not coded in ASCII.

Figure 11.11 illustrates the various formats for TAG and length indicators.

TAG	Length	G	S	M
		ASCII ↓	ASCII ↓	ASCII ↓
04	03	47	53	4D

Figure 11.10 Format of octetstring.

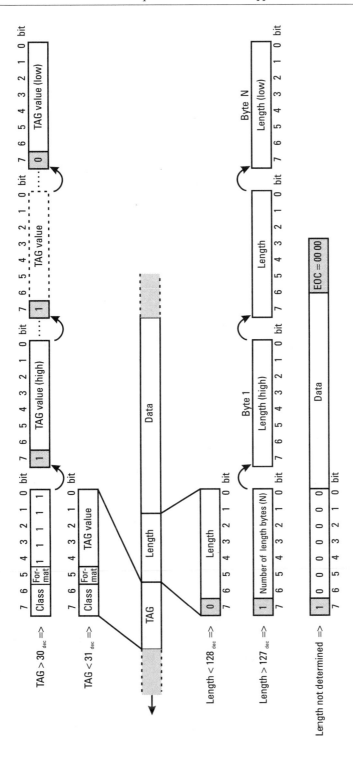

Figure 11.11 Various formats for TAG and length indicators in TCAP. Note that bit 0 is always sent first, despite the information about the direction, right to left.

11.1.3.4 Presentation of the Length Indicator

Problems similar to those described for the coding of the TAG arise in the coding of the length of a message or a parameter. The theoretical limit for coding the length field with just 1 byte is 255 bytes. For all practical purposes, that value would be suitable for the time being, because SCCP UDT messages are limited to 255 bytes. However, to be safe in the future and allow for additional applications, a solution was needed that allowed the coding of a field of any necessary (arbitrary) length.

Another requirement was that fields of undefined length be allowed, at least for constructor parameters, where MAP possibly does not know the actual length. Therefore, three different representations needed to be distinguished (see Figure 11.11).

- For "small" length (less than 128 bytes), the length indicator field is 1 byte long. Only bits 0 through 6 are used, which allows coding of values between 0 and 127_{dez} (0111 1111_{bin}). Bit 7 is always zero; hence, the limit of 255 cannot be reached.
- For a "large" length (greater than 127 bytes), the length indicator field needs to be extended by the necessary number of bytes. For that purpose, bit 7 is set to 1 and bits 0 through 6 then indicate the number of bytes to follow, which carry length information. For example, when the first byte of the length indicator field is coded as 1000 0111_{bin}, then 7 bytes ($111_{bin} = 7$) of length information will follow, for a total of 8 bytes of length information.
- For an undefined length, the length indicator field is 1 byte long and is fix-coded with 80_{hex}. Here, bits 0 through 6 are all of 0 value and bit 7 is set to 1. However, the end of a parameter with undefined length needs to be indicated, too. For that purpose, a special end mark, the end of contents (EOC), is added. The EOC consists of 2 bytes, coded with all zeroes. Note that an undefined length indication may be used only for constructor parameters.

11.1.3.5 "Large" TAG and Length Indicator

In this example, a parameter needs to be coded for the transmission via TCAP. TAG and length are as follows:

$$\text{TAG type} = 2222_{dez} = 08AE_{hex} = 0000\ 1000\ 1010\ 1110_{bin}$$

$$\text{length} = 3333_{dez} = 0D05_{hex} = 0000\ 1101\ 0000\ 0101_{bin}$$

This represents a parameter of class "constructor" with the format "context specific." In both cases, 1 byte is not enough to code the particular fields. The value for TAG is greater than 30 and the length is greater than 127. For that reason the expanded form has to be applied.

The TAG

Three bytes are necessary to represent the TAG, as shown in Figure 11.12(a). Byte 1 is used to define class and format, while bytes 2 and 3 contain the actual TAG value.

- The value for class is 10_{bin} = context specific.
- The value for format (F) is 1_{bin} = constructor.
- The value for TAG of the first byte is 11111_{bin} and indicates that the TAG is expanded.
- The actual TAG value of $0000\ 1000\ 1010\ 1110_{bin}$ needs to be coded within bytes 2 and 3 and then be inserted right-aligned. Bit 7 may not be used and leading zeros are omitted.

Coding of the type TAG, therefore, is BF 91 $2E_{hex}$.

The Length Indicator

The length indicator itself requires 2 bytes ($0D05_{hex}$). Together with the first byte (to indicate an extended length field), that totals 3 bytes, as shown in Figure 11.12(b):

Byte 1			Byte 2			Byte 3		
Class F	TAG extended	E	TAG value (high)	E	TAG value (low)
bit 7 6 5 4	bit 3 2 1 0	bit 7	6 5 4 3 2 1 0	bit 7	6 5 4 3 2 1 0
1 0 1 1	1 1 1 1	1	X X X X X X X	0	X X X X X X X
1 0 1 1	1 1 1 1	1	0 0 1 0 0 0 1	0	0 1 0 1 1 1 0
B	F		9		1		2	E

11.12(a) Coding a large TAG.

Byte 1	Byte 2	Byte 3
Number of subsequent bytes	Length indicator (high)	Length indicator (low)
E bit	bit	bit
7 6 5 4 3 2 1 0	7 6 5 4 3 2 1 0	7 6 5 4 3 2 1 0
1 0 0 0 0 0 1 0	x x x x x x x x	x x x x x x x x
1 0 0 0 0 0 1 0	0 0 0 0 1 1 0 1	0 0 0 0 0 1 0 1
8	2	0D	05

Figure 11.12(b) Coding a large length indication.

- Bit seven of the first byte is the extension mark and needs to be set to 1.
- Because the actual length requires 2 additional bytes, bits 0–6 have to be coded with a value of $000\ 0010_{bin} = 2$.
- Now, the value for the length ($0D05_{hex}$) is inserted into bytes 2 and 3, starting from the right.

Coding of the length, therefore, is $82\ 0D\ 05_{hex}$.

11.1.4 TCAP Messages Used in GSM

ITU-T, in its Recommendations Q.772 and Q.773, has defined five TCAP messages, of which GSM uses four. The four messages used in GSM are illustrated in Figure 11.13. The white areas in Figure 11.13 are optional parts; the mandatory parts are shaded.

Table 11.2 provides details on the various TCAP messages. Note that the messages form only the transport container for the MAP content.

11.1.4.1 The Dialog Portion

The dialog control of TCAP messages was mentioned at the beginning of this chapter. The dialog control is, together with some further information, part of the optional dialog portion of a TCAP message (see Figure 11.15).

Structured Versus Unstructured Dialog

The term *dialog portion* has a meaning different from that of the term *dialog*. A dialog refers to the whole communication process of exchanging information between users. One has to distinguish between an unstructured dialog and a structured dialog. GSM uses the services of the structured dialog only. For that

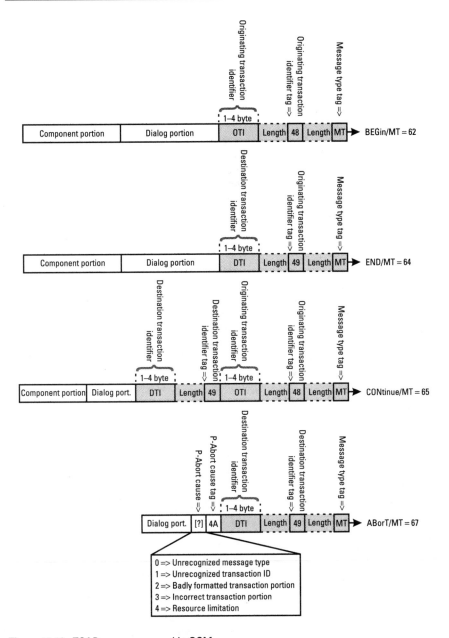

Figure 11.13 TCAP messages used in GSM.

reason, unstructured dialog will not be dealt with in more detail. However, it should be mentioned that unstructured dialog, when used, is transmitted in a

Table 11.2
TCAP Messages Used in GSM

ID (Hex)	Name	Abbr.	Interface	Description
62	BEGin	BEG	all NSS	By sending a BEG message, TCAP opens a dialog for one user (this is MAP in the case of GSM) to another user. The BEG message comprises the originating transaction identifier, which identifies a dialog within TCAP, or more precisely, within the transaction layer. Additionally, the invoke ID of the optional component part may be used to identify the dialog.
64	END	END	all NSS	An END message is sent specifically, when a process needs to be terminated, which was started by a BEG message. An END message may be the direct response to a BEG message. The END message also has an optional component part, which may contain MAP data.
65	CONtinue	CON	all NSS	The CON message is used between the beginning and ending of a process in order to transport information between MAP users. A CON message comprises both an originating transaction identifier as well as a destination transaction identifier. The first CON message, which is sent after a BEG message, confirms that the requested protocol and application context are accepted.
67	ABorT	ABT	all NSS	Both TCAP and MAP may use the ABT message to terminate a process at any time if an error occurs or if a request cannot be processed. A reason for termination may or may not be provided. A distinction is made, however, between the source for termination. When the *service provider*, which is TCAP, initiates the termination, *P-ABORT* is used, when the *user*, which is MAP initiates the termination, U-ABORT is used. For the first category, the ABT message provides a P-Abort-Cause field. In the second case, the reason for abortion is sent within the field *Dialog Control*. A frequent reason for the abortion of a process is the incompatibility between the protocol versions of MAP (application context name) in the two peers of a dialog. One side, for example, requests an old or a new protocol, which the partner does no longer or not yet support.

UNI message. The difference between the two dialog types is that unstructured dialog actually is not a real dialog but a one-time data transmission, in which the sender does not expect any feedback or answer. In contrast, structured dialog consists of the beginning, the execution, and the termination of a communication process between two peers, as illustrated in Figure 11.14.

A user opens a structured dialog by sending a BEG message. The BEG message identifies a transaction on the serving side by means of the originating transaction identifier. If the addressee is able to accept the dialog, then it answers with a CON message, which contains both an originating transaction identifier and a destination transaction identifier. Following that, both sides may continue to send additional CON messages. To end a dialog, one side sends an END message.

In the case of a very short dialog, the process consists only of a BEG message and an END message.

The Dialog Unit

This optional area in a TCAP message is used to allow both end points of a connection to synchronize processing of the data, contained in the component part. Three different dialog units have been defined as part of the dialog portion:

- The dialog request: proposal of a protocol;
- The dialog response: confirmation of the proposed protocol;
- The dialog abort: termination of a dialog (may or may not be related to the proposed protocol).

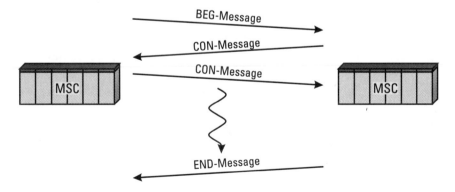

Figure 11.14 The structured dialog in an MSC-to-MSC transaction.

Figures 11.15(a) and 11.15(b) show the formatting of the dialog request, dialog response, and dialog abort. Exactly one dialog unit may be present per TCAP message.

The dialog request can be compared to a negotiation between two individuals about which language their conversation will be in. TCAP uses the application context name for those discussions that contain an identifier for the standard and the protocol that will be used for a transaction.

In GSM, the string

> *Ccitt-identified-organization.etsi.mobileDomain.gsm-Network.ac-id Application Context Name. Version*

tells the recipient exactly which MAP module in what version is proposed by the sender. Every parameter of that string is represented in a TCAP message by exactly 1 byte. In this context, *parameter* means a part of the string that lies between two dots. Depending on the position and the value of a byte, the meaning of the value, for example, 01 for the fourth byte is gsm-Network. For the GSM-MAP, only the last three bytes are actually usable:

- Byte 5: ac-id application context identification (always the same);
- Byte 6: value for ac-id (assigned by GSM);
- Byte 7: version number of ac-id (e.g., Phase 1 or Phase 2).

For example, in the context of a location update or an IMSI detach, the application context [networkLocUp V2] is used by the VLR for communication with the HLR. That corresponds to the byte sequence 01 02 (examined in more detail in Chapter 12). The information is sent to the HLR, together with the leading five bytes, which are always coded with a fixed value. The received application context name allows the HLR to derive the information, which software module needs to be used, or how the data within the component portion shall be treated.

As already indicated, TCAP and MAP follow the ASN.1 when coding data. The same applies for the dialog unit. That explains terms like *object descriptor* and *external*, the values of which are listed in the Glossary entry *ASN.1*.

Transaction Capabilities and Mobile Application Part 203

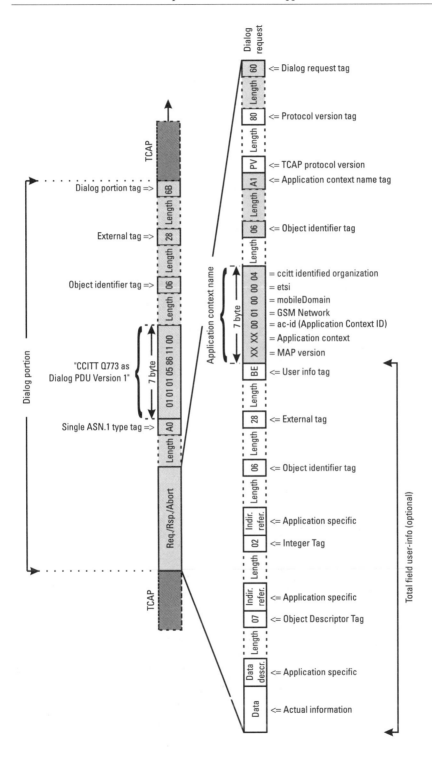

Figure 11.15(a) The dialog portion (part 1).

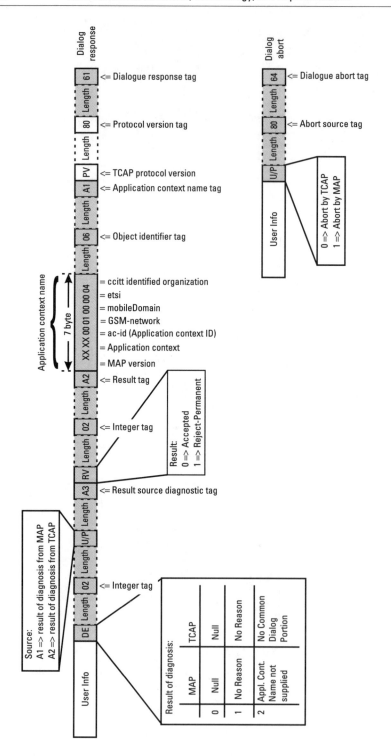

Figure 11.15(b) The dialog portion (part 2).

11.1.4.2 The Component Portion

The component portion, if present, contains the actual user data. This is MAP signaling information in the case of GSM. Different units have been defined for the component portion, just as for the dialog portion. Figures 11.16(a) and 11.16(b) describe these components.

Invoke Component

The invoke component starts an operation on the receiver side. An operation is defined by the operation code and is application specific. In GSM, the MAP operation codes correspond to these operation codes (see Section 11.2.3). Similar to the message types in other signaling standards like SS7, these values indicate the composition of the parameter field that follows.

Every component contains an invoke ID, to allow for different components to be unambiguously dedicated to an operation or a dialog. The invoke component also may contain a linked ID, to allow relations between different dialogs. The linked ID contains the value of the invoke ID of the second dialog, which is linked to the first.

The MAP user data are carried in the optional parameter field of an invoke component.

Return Result Component

The return result component (*last* and *not last*) is the result of a dialog opened by an invoke component and is transported in a return result component. Note that because of the limited capacity of UDT messages the transmission of result data might be segmented. In that case, the return result (*not last*) component is used for all data segments but the very last one (as shown in Figure 11.17). Note that the last result data segment is always transferred in a return result (*last*) component. Of course, that also applies if no segmentation was necessary.

Return Error Component

A return error component is the answer to an invoke component if an operation cannot be completed successfully. "Not successful" does not necessarily mean some sort of protocol error; it could refer to other problems, like a subscriber not responding to paging or that a particular IMSI is unknown in the HLR. Figure 11.16(b) shows examples of local error codes. Furthermore, the example in Section 11.1.4.3 shows the decoding of a TCAP message that contains a return error component.

206   GSM Networks: Protocols, Terminology, and Implementation

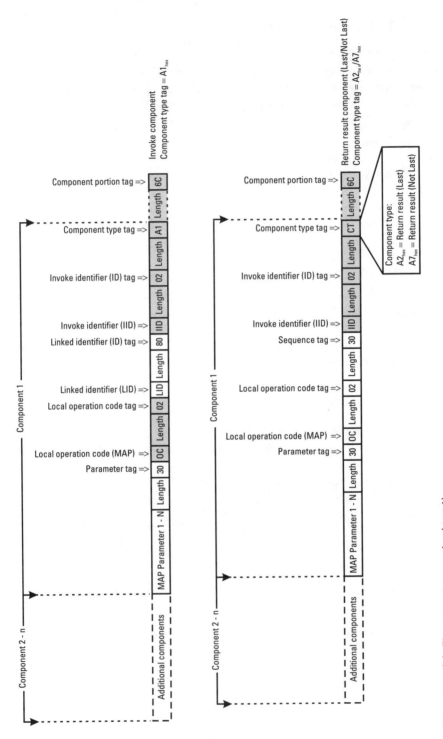

Figure 11.16(a) The component portion (part 1).

Transaction Capabilities and Mobile Application Part

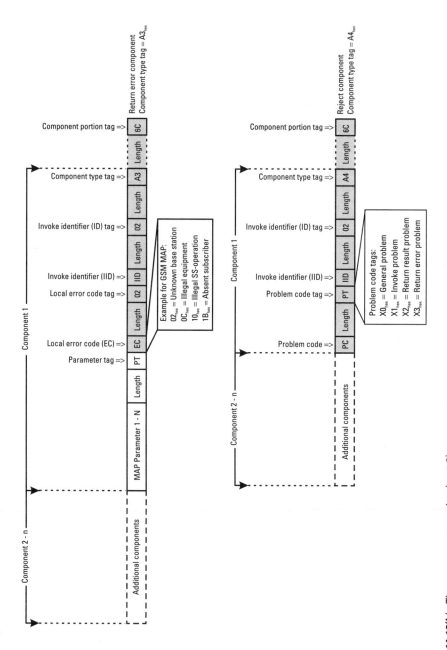

Figure 11.16(b) The component portion (part 2).

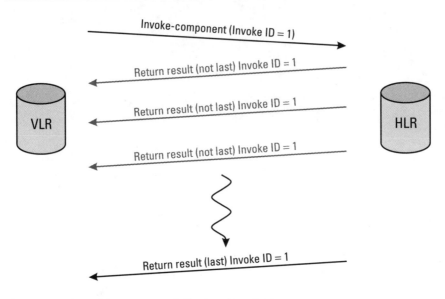

Figure 11.17 Use of the return result (*last*) and (*not last*) components.

Reject Component

If an application is unable to process a component because of errors, the problem is conveyed to the sender by a reject component. With the exception of the reject component itself, every component may be rejected.

The available error causes are categorized into four groups, as listed in Table 11.3. The numbers in the table correspond to the Problem Code (PC) in Figure 11.16(b).

11.1.4.3 Decoding of an END Message With Dialog and Component Portions

Figure 11.18 shows an extract of a trace file that was captured by a protocol tester. The trace file presents the TCAP content of an END message, sent from a VLR to an HLR. Although a return error is reported, there is no real error, simply the feedback that paging for a subscriber was unsuccessful. This example also illustrates how TAG and length indications affect the size of a message.

11.2 Mobile Application Part

The MAP was mentioned frequently in the first part of this chapter. The reason for this is the tight cooperation between MAP and TCAP in the NSS. This second section focuses on the processes and procedures within MAP itself; the

Table 11.3
Problem Codes for the Reject Component

General problem (X0):	
0	Unrecognized component
1	Mistyped component
2	Badly structured component
Invoke problem (X1)	
0	Duplicate invoke ID
1	Unrecognized operation
2	Mistyped parameter
3	Resource limitation
4	Initiating release
5	Unrecognized linked ID
6	Linked response unexpected
7	Unexpected linked operation
Return result problem (X2)	
0	Unrecognized invoke ID
1	Return result unexpected
2	Mistyped parameter
Return error problem (X3)	
0	Unrecognized invoke ID
1	Return error unexpected
2	Unrecognized error
3	Unexpected error
4	Mistyped parameter

emphasis will be on MAP as the interface or adapter between TCAP and the application.

11.2.1 Communication Between MAP and its Users

All the signaling protocols introduced so far, from LAPD through TCAP, are peer-to-peer protocols, that is, horizontal protocols, in the way the OSI

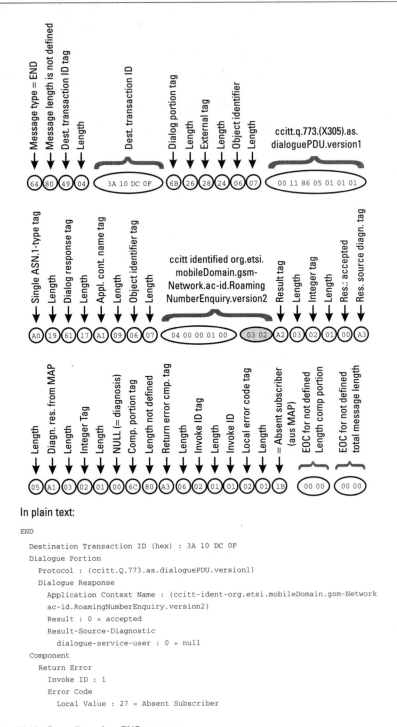

Figure 11.18 Decoding of an END message.

Reference Model considers them. The primitives, as the carrier of vertical data exchange, were neglected. Nevertheless, LAPD, SCCP, and TCAP all need primitives to communicate with the adjacent higher or lower layer. For the reader's comprehension, these primitives were of less importance. The situation is different for the communication between MAP and TCAP, so this is the time to discuss primitives in more detail. Figure 11.19 illustrates the relevant part of the OSI Reference Model.

Because the application itself (HLR, VLR, etc.) is not part of the OSI Reference Model, so-called MAP services are required for control tasks and data exchange between the different applications and MAP. The MAP services actually are primitives. Let's use the expression *MAP application* as a collective term for all involved GSM subsystems, from the MSC to the EIR.

11.2.2 MAP Services

The communication between MAP and an application is done via MAP services, in which we have to distinguish between common MAP services for pure communication control between MAP and application and special MAP services as the carrier of signaling data. Both variants are dealt with in the following sections.

11.2.2.1 Direction Dependency of MAP Services

The tasks of Layers 1 through 7 of the OSI Reference Model are transparent for the MAP application; they can be considered as some kind of black box. The MAP application sees only the communication with MAP, which is performed by MAP services. MAP in Layer 7 receives commands and answers from the application that are conveyed to the peer entity via TCAP and the remaining layers of the OSI Reference Model. On the other hand, MAP receives commands and answers from TCAP (Layer 6) that actually come from the peer

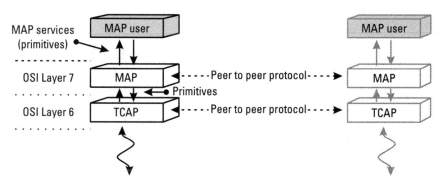

Figure 11.19 MAP in the OSI Reference Model.

entity and need to be forwarded to the application. It is important for MAP and for the MAP applications to distinguish not only between the various MAP services but also between the two possible directions of those services. For that reason, up to four different variants were defined for every MAP service (as for primitives in general). An overview is provided in Figure 11.20. Let the initiating MAP application be A and the responding MAP application be B. A sends a *request*, which translates into an *indication* on the side of MAP application B. B's answer is sent back in a *response* and translated into a *confirmation* from MAP to A. Although the differentiation is of little significance during protocol testing, it is important for a thorough comprehension of MAP.

11.2.2.2 Common MAP Services

Six common MAP services can be used to control a communication between MAP and its application. Depending on the task of the MAP service, all four or only some of the primitives—*Request, Indication, Response,* and *Confirmation*—are needed.

MAP-DELIMITER Service

By sending this primitive, the application indicates that a data packet is complete and ready to be passed to the peer entity. Such a data packet may contain a MAP-OPEN service for communication control, special MAP services (with signaling data), or both. Only the *Request* and *Indication* variants are defined for the MAP-DELIMITER service.

MAP-OPEN Service

By means of this primitive, an application requests MAP to establish a dialog with another application. The MAP-OPEN service includes the specification of the requested transaction (application context name) and identifies sender and addressee. Neither parameters nor data are included. All four primitives—*Request, Indication, Response,* and *Confirmation*—are defined for the MAP-OPEN service.

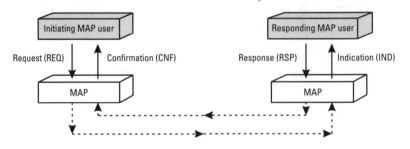

Figure 11.20 Direction dependency of MAP services.

MAP-CLOSE Service

The MAP-CLOSE service is used to terminate an existing process. The primitive is passed to the MAP and then forwarded to TCAP when a MAP application intends to terminate (not interrupt) a dialog. Only the *Request* and *Indication* variants are defined for the MAP-CLOSE service.

MAP-U-ABORT Service

The abbreviation stands for MAP User Abort and indicates that an application wants to interrupt a dialog. Only the *Request* and *Indication* variants are defined for the MAP-U-ABORT service.

MAP-P-ABORT Service

The abbreviation stands for MAP Service Provider Abort and indicates that TCAP wants to interrupt or has already interrupted a dialog. Only the *Indication* variant is defined for the MAP-P-ABORT service.

MAP-NOTICE Service

The MAP-NOTICE service provides an application with information about problems on the peer side. Only the *Indication* variant is defined for the MAP-NOTICE service. In particular, when a TCAP message with a reject component and specific problem codes is received, the MAP application gets a MAP-NOTICE service. Reasons might be protocol errors (e.g., duplicate invoke ID) or unexpected data values and parameter types.

11.2.2.3 Special MAP Services

The purpose of the common MAP services is to control the communication between MAP and its applications. Although the MAP-OPEN service already contains the application context name and, hence, the requested protocol for the dialog to be established, only a special MAP service such as *updateLocation* contains the actual parameters. Like the message type for other signaling standards, the local operation codes identify the special MAP services within MAP. As for all primitives, up to four variants (Request, Indication, Response, and Confirmation) are defined for the special MAP services. Two examples are the following:

- The local operation code *forwardAccessSignaling* is a service for transparent transmission of BSSAP data between MSCs after an inter-MSC handover. This service is nonconfirmed, that is, no acknowledgment is returned when a *forwardAccessSignaling* code is received. Therefore, only the Request and Indication variants are necessary for the *forwardAccessSignaling* service.

- The local operation code *updateLocation* is required directly after a location update for the new VLR to update the location information in the HLR. Because this is a confirmed service, it requires all four variants: Request, Indication, Response, and Confirmation.

The strange representation style of the local operation codes becomes obvious in the preceding examples. The words are concatenated without any blanks, and the first letter of the first word is in lowercase, while the first letters of all other words are uppercase. That is the defined way to present local operation codes.

11.2.3 Local Operation Codes of the Mobile Application Part

This section presents the special MAP services (local operation codes) as they are defined for GSM Phase 2. Note that GSM does not consider the B-interface, that is, the interface between MSC and VLR, to be an external interface. Therefore, no binding recommendations were made for that interface nor for the information that needs to be sent over it. For that reason, this section contains no local operation codes that are used exclusively on the B-interface.

Table 11.4 explains the tasks and the meanings of the various MAP local operation codes. The information provided in the third column indicates which interfaces the respective local operation code have used. (The Glossary and Chapter 4 provide definitions of the interfaces.)

Table 11.4
MAP Local Operation Codes

Op.-Code (Hex)	Name	Interface	Description
02	updateLocation	D	When a MS has successfully performed a location update in a new VLR area, this message is used to inform the HLR. If several MSCs are served by one VLR, then this message is also sent when the MS changes the MSC within the VLR area.
03	cancelLocation	D	When a MS roams into a new VLR area this message is used to inform the old VLR that it will delete the related subscriber data.

Table 11.4 (continued)

Op.-Code (Hex)	Name	Interface	Description
04	provideRoamingNumber	D	Is used between VLR and HLR in case of an MTC (mobile terminating call) to provide routing information to the HLR. This information, the roaming number (MSRN), is then sent to the Gateway-MSC (see also local operation code 16: sendRoutingInfo).
07	insertSubscriberData	D	The HLR uses this operation code to provide the VLR with the current subscriber profile, e.g., during a location update procedure.
08	deleteSubscriberData	D	The HLR uses this message to inform the VLR that a service (in particular a supplementary service) was removed from a subscriber profile.
0A	registerSS	B, D	Is used to enter supplementary service data for a specific subscriber. This may be accompanied by an automatic activation of this supplementary service, e.g., when the subscriber enters a call forwarding number. In this case the supplementary service is automatically activated at the same time. For the MSC and the VLR, the respective data is transparent. This data is only relevant for the HLR.
0B	eraseSS	B, D	This local operation code is used to delete supplementary service data for a specific subscriber that was previously entered by registerSS.
0C	activateSS	B, D	By this local operation code a supplementary service is switched on for a specific subscriber. Supplementary services to be activated by the subscriber, are, for example, CLIP and CLIR. Like the other SS messages, this message is transparently passed from the MSC to the VLR and then to the HLR.
0D	deactivateSS	B, D	This "turns off" a Supplementary service that was previously activated. It is the reverse operation to activateSS.
0E	interrogateSS	B, D	This local operation code allows to query the state and the details regarding a Supplementary service in the HLR for a specific subscriber. With one interrogateSS message exactly one Supplementary service can be queried.

Table 11.4 (continued)

Op.-Code (Hex)	Name	Interface	Description
11	registerPassword	B/D	Is used to enter or modify a password for a Supplementary service. After the HLR has received this message, it responds with getPassword messages in order to request the old password, the new password, and to verify the new password. This operation is blocked if the old password was incorrectly entered for three consecutive times.
12	getPassword	B/D	If a subscriber wants to change the current password or modify or activate a Supplementary service, which is password protected, then the HLR requests this password in a getPassword message. This operation is blocked if the old password is incorrectly entered for three consecutive times.
16	sendRoutingInfo	C	The Gateway-MSC sends this message in case of an MTC (mobile terminating call) to the HLR of the called subscriber in order to obtain routing information. This routing information consists of the MSRN (mobile station roaming number), which is formatted like an ordinary telephone number (see in Glossary). When the HLR receives a sendRoutingInfo request it sends a provideRoamingNumber request to the VLR in which area the respective subscriber is currently roaming.
1D	sendEndSignal	E	The sendEndSignal request is used after a inter-MSC handover. After a successful handover from MSC A to MSC B, MSC B sends a sendEndSignal request to MSC A, which allows MSC A to release the radio resources. If MSC A is the MSC where the call was originally established, then it keeps the overall call control even after the inter-MSC handover. In this case, MSC A is referred to as "anchor MSC." Consequently, MSC B does not receive any information on when that call is released, i.e., when MSC B may trigger the release of its own radio resources. To cope with this situation, MSC B receives the respective information in the sendEndSignal response from MSC A.

Table 11.4 (continued)

Op.-Code (Hex)	Name	Interface	Description
21	ProcessAccess Signaling	E	An anchor-MSC keeps control of a call, even after successful handover from MSC A to MSC B. In order to establish a transparent connection between MSC A and MS, the local operation codes processAccessSignaling and forwardAccessSignaling were defined. Their task is to transfer BSSAP messages between MS and MSC A, i.e., on the path MS MSC B MSC A. The difference between the two operation codes is the direction. ProcessAccessSignaling is sent from MSC B to MSC A, hence, it carries data from the MS to MSC A, while forwardAccessSignaling is used in the reverse direction, i.e., it carries data from MSC A towards the MS.
22	ForwardAccess Signalng	E	Please refer to processAccessSignaling.
25	reset	D	Reset is only used when a HLR is brought back to service after an outage. The HLR sends this local operation code to all VLR's, in which mobile stations of that HLR are registered according to the data that is still available after the outage.
26	Forwardcheck SS-Indication	B / D	ForwardcheckSS-Indication is optional and sent to all affected mobile stations after an HLR outage. The subscriber is requested to synchronize its SS data with the network.
2B	checkIMEI	F	This local operation code is used to convey an IMEI (international mobile equipment identity) between MSC/VLR and EIR.
2D	sendRoutingInfo-ForSM	C	This local operation code is used by the SMS-Gateway-MSC during the MT-SMS procedure (mobile terminating SMS) in order to deliver a short message to the MSC as to which area the subscriber roams. This request for routing information from the SMS Gateway MSC contains the MSISDN of the subscriber, while the result contains the ISDN number (routing address) of the destination MSC. This address is used by the SCCP to forward the short message in a forwardSM message.

Table 11.4 (continued)

Op.-Code (Hex)	Name	Interface	Description
2E	forwardSM	E	The MO-SMS procedure (mobile originating SMS) as well as the MT-SMS procedure (mobile terminating SMS) is used in both cases to carry a short message between the MSC where the subscriber roams and the MSC, which has a connection to the SMS Service Center (= SMS Interworking MSC).
2F	reportSM-DeliveryStatus	C	When the transmission of a short message from the SMS Service Center to the MS was unsuccessful, (e.g., because the subscriber was not reachable) then the MSC returns a negative response to the SMS Gateway MSC. In this case, the SMS Gateway MSC sends a reportSM-DeliveryStatus to the HLR to allow for a later delivery of the short message. The HLR sets a message waiting flag in the subscriber data of the subscriber, sends an alertServiceCentre message (as an information about the negative result of the short message transfer) to the SMS Interworking MSC, and waits until the subscriber is reachable again. When the VLR, which is also aware of the unsuccessful SM delivery, detects that the subscriber is reachable again it sends a readyforSM message. When the HLR receives this message it reacts with an alertServiceCentre message to the SMS Interworking MSC, which in turn informs the SMS Service Center. Now the delivery process to the MS can start again with a forwardSM message.
32	activateTraceMode	D	Is used by the HLR to activate the trace mode for a particular subscriber (IMSI). Note that the request for subscriber tracking is not originated by the HLR but the OMC. After receiving an activateTraceMode request, the VLR waits until that particular MS becomes active. When this is the case, the VLR sends an internal request to the MSC to trace the MS. In contrast to most other MAP operation codes, no acknowledgment is expected in case of activateTraceMode. The results are directly passed from the BSS/MSC to the OMC.
33	deactivateTraceMode	D	When the HLR receives this operation code it turns the active trace mode off (see activateTraceMode).

Table 11.4 (continued)

Op.-Code (Hex)	Name	Interface	Description
37	sendIdentification	G	When an MS changes the VLR area the new VLR queries the old VLR with sendIdentification for the currently valid authentication data. If the new VLR is unable to identify the old VLR, e.g., when the PLMN is changed as well, then this information can also be retrieved from the HLR, in this case by means of sendAuthenticationInfo.
38	sendAuthenticationInfo	D	Please refer to sendIdentification.
39	restoreData	D	When a VLR receives a provideRoamingNumber request from the HLR for a) an IMSI, unknown in the VLR or b) an IMSI, where the VLR entry is unreliable after an HLR outage, then the VLR sends a restoreData request to the HLR, in order to synchronize the data between VLR and HLR.
3A	SendIMSI	D	When the VLR receives a request from the OMC to identify a subscriber, based on his MSISDN (directory number) then this request is handled by the exchange of sendIMSI operation codes between VLR and HLR.
3B	processUnstructuredSS Request	B/D	This operation code is used to provide means for the handling of additional, non-GSM standardized supplementary services within a PLMN (the unstructured supplementary services). Unlike unstructuredSS-Request, processUnstructuredSS Request is used by both sides, the MS and the addressed NSS entity, if the MS has initiated the transaction.
3C	unstructuredSS-Request	B/D	This operation code is used to provide means for the handling of additional, non-GSM standardized supplementary services within a PLMN (the so-called unstructured supplementary services). unstructuredSSRequest is used by both sides, if the request was initiated by the NSS.
3D	unstructuredSS-Notify	B/D	Unlike processUnstructuredSSRequest and unstructuredSSRequest, unstructuredSSNotify is used when an additional feature in the NSS needs to transport USSD (unstructured supplementary services data) to the MS, without expecting an acknowledgment from the subscriber. The MS, however, confirms receipt of the corresponding data by returning an empty unstructuredSSNotify (FACILITY message on the Air-Interface).

Table 11.4 (continued)

Op.-Code (Hex)	Name	Interface	Description
3F	informService-Centre	C	The HLR sends this operation code to the SMS Gateway MSC, when a sendRoutingInfoForSM was received for a subscriber who is currently not available.
40	AlertService-Centre	C	Please refer to reportSM-DeliveryStatus
42	readyForSM	D	Please refer to reportSM-DeliveryStatus
43	purgeMS	D	If a MS was inactive for an extended period of time, that is, no call or location update was performed, then the VLR sends a purgeMS request to the HLR. This indicates that the VLR has deleted the data for that particular MS. The HLR sets the "MS Purged" flag and no longer attempts to reach the MS in case of an incoming call.
44	prepareHandover	E	At the beginning of a inter-MSC handover (MSC A MSC B) a prepareHandover request and response is sent between both MSCs in order to exchange BSSAP messages and to trigger the activation of a TCH in MSC B. The prepareHandover message is used in particular to transport the handover number and the two BSSMAP messages: HND_REQ and HND_REQ_ACK.
45	prepareSubsequent Handover	E	If after an inter-MSC handover another inter-MSC handover should become necessary, either back to MSC A or to a third MSC (MSC C), then MSC B sends a prepareSubsequentHandover message to MSC A. This message contains all necessary information for MSC A to send a prepareHandover message to MSC C.

11.2.4 Communication Between Application, MAP, and TCAP

An application communicates with MAP by means of common MAP services and special MAP services. But how does MAP pass these services to TCAP? And how does TCAP pass information it receives, commands, and responses to MAP? This section provides answers to those questions.

The term *dialog* stems from the vocabulary of TCAP and addresses the exchange of data between two TCAP users. GSM uses only the services of the so-called structured dialog, which is used when, upon delivery of data, a reaction, an acknowledgment, or an answer is expected from the recipient.

With respect to data transmission between MAP, TCAP, and application, this restriction simplifies the situation, since a dialog between MAP applications always has to be structured. That requires, from the perspective of TCAP, that it starts with a BEGIN message and, in case of no errors, terminates with an END message. A special case is the abortion of a dialog with an ABORT message, which can be sent by either MAP or TCAP.

Figure 11.21 illustrates, by means of the example of the cancelLocation service, how the MAP services are applied internally. Note that the primitives between MAP and TCAP are not shown. The cancelLocation service is required after a location update if a MS has changed the VLR area and the subscriber data in the old VLR need to be deleted.

- The application, in this case, the HLR, transfers a MAP-OPEN service REQ to MAP. It contains the ISDN addresses of VLR and HLR which are required for addressing by the SCCP. Furthermore, the MAP-OPEN service REQ contains the requested application context name for the dialog, in this case, [LocationCancel.version2] = [2.2]. This application context is required for the TCAP dialog portion.
- The message in the frame with double lines shows the special MAP cancelLocation service REQ which carries the actual signaling data, which in this case is the invoke ID, IMSI, and the local mobile subscriber identity (LMSI). These data are transported in the component portion of the subsequent TCAP message.
- The application requests MAP, by means of the MAP-DELIMITER service REQ, to pass all the information to TCAP.
- In this situation, TCAP will send a BEG message to the requested address that includes the corresponding dialog and component portions.
- The TCAP entity on the side of the VLR receives a BEG message after SS7 and SCCP have properly routed that message to the VLR, and passes the information in a primitive to MAP. If MAP does not know this specific application context, it sends an ABORT primitive back to TCAP. In such cases, the application in the VLR does not receive any indication.
- In the positive case, if the VLR supports the application context then MAP passes the address information and the application context in a MAP-OPEN IND to the application, in this case, to the VLR. The application context allows the VLR to conclude how the received data have to be processed. At this stage, the VLR does not yet know the identity of the subscriber whose entry is to be deleted.

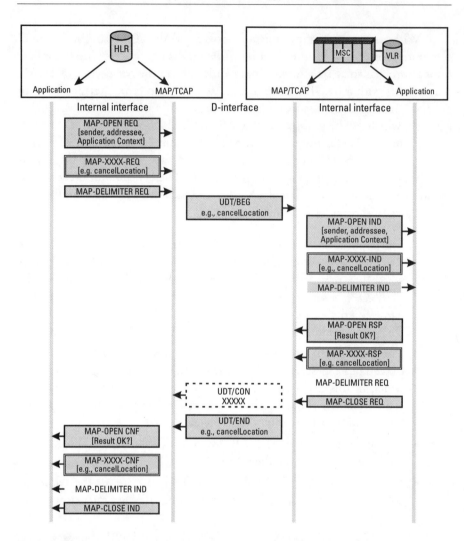

Figure 11.21 Interaction of application, MAP, and TCAP during cancelLocation.

- Exactly this information is provided by the MAP cancelLocation service IND, which corresponds to the content of the component portion of the TCAP BEG message. In Figure 11.21, the primitive is presented in a double-lined frame.
- When the VLR has received and evaluated all necessary information, it deletes the corresponding subscriber data, including the LMSI. Then the VLR responds to MAP in a MAP-OPEN RSP, which contains the information as to whether the result was positive

or negative but not the address information previously received in MAP-OPEN IND.
- The invoke ID is included in the MAP cancelLocation service RSP, sent to MAP.
- In the scenario in Figure 11.21, the next message sent to MAP is the MAP-DELIMITER REQ. This message is sent, however, only when a dialog should not be closed but continued. A MAP-CLOSE REQ is sent in the case of the cancelLocation process. Thus, the MAP-DELIMITER REQ is marked as optional.
- Now, the difference between closing and continuing a dialog becomes more obvious. TCAP would use a CON message, to respond to the VLR in case the dialog needs to be continued (if MAP-DELIMITER REQ was received). In case of cancelLocation or as a reaction or response to the MAP-CLOSE REQ, TCAP sends an END message back to the HLR.
- You might wonder at this point if the confirmation for the opening of the dialog is still pending in the HLR. That is taken care of by sending the MAP-OPEN CNF from MAP to the HLR, after the TCAP-END message is received. The same applies for the special MAP service cancelLocation.
- Receipt of MAP-CLOSE IND terminates the dialog on both sides.

12

Scenarios

This chapter applies the acquired knowledge base of the previous chapters to describe the various GSM subsystems via signaling protocols. Every presented scenario is explained in detail.

Before presenting those details, however, the commonality, or "red thread" of the scenarios should be emphasized. For that purpose, the block diagram in Figure 12.1 applies to all: MOC (mobile originating call), MTC (mobile terminating call), and LU (location update). The following is an explanation of the individual blocks in Figure 12.1:

- Only in a MTC does the network search for a particular subscriber (paging).
- When the MS is located or when the MS initiates a call, a control channel between MS and BSC has to be established.
- The MS uses the control channel for identification and indicates to the BSC in detail which service is requested.
- The BSC passes the service request of the MS to the NSS. For that purpose, the BSS has to request an SCCP connection from the MSC.
- The NSS reacts on a connection request of any kind with a request for authentication (except for an emergency call). Additionally, the IMEI may be checked.
- Ciphering between BTS and MS is activated in successful authentication. Ciphering prevents tapping into the Air-interface.
- Additional information between MS and NSS are exchanged after activation of ciphering. The additional information either terminates

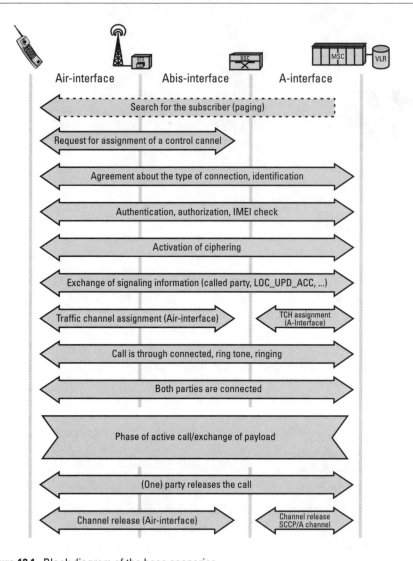

Figure 12.1 Block diagram of the base scenarios.

a successful LU, or, in case of a connection request, defines the details of that connection (e.g., directory number of the called subscriber, required technical capabilities of the network and the MS). The process is synchronized between MS and NSS.

- The assignment of the TCH on the A-interface and Air-interface is done separately, except in the case of off-air call setup (OACSU). Up to this point, the communication has been done via a control channel.

- The system waits, after assignment of the traffic channel, until an end-to-end connection is in place. At the end of this phase, the telephone on one side rings, and the other side hears the ring-back tone.
- When the called subscriber takes the call, the actual conversation begins and charges apply from then on.
- Both ends terminate the call after the conversation has ended. This is the trigger for the MSC, as well as for the MS, to release all the occupied channels and resources.

12.1 Location Update

12.1.1 Location Update in the BSS

An MS performs LU on several occasions: every time it changes the location area, periodically, when a periodic location update is active, or with IMSI attach/detach switched on at the time when it is subsequently turned on again. The only subsystems shown in Figure 12.2 are the MSC/VLR, BSC, BTS, and MS. Nevertheless, if the VLR area changes, the HLR, as well as the old VLR, are involved, too. Furthermore, if the equipment-check is active, the EIR is also involved. Figure 12.3 shows LU from the perspective of the NSS.

12.1.2 Location Update in the NSS

Figure 12.3 shows a LU in which the VLR changes. In this case, the HLR is particularly involved in the overall process. When the LU involves no VLR change, the HLR does not need to be accessed. The HLR only has information about the VLR area of a subscriber; it has no information about the details of the location area. If the equipment check is turned on, the EIR is checked with every activity of an MS.

12.2 Equipment Check

The GSM standard enables a network operator to not only verify the identity of a subscriber by means of authentication but also to check the mobile equipment (ME) as such, which is identified by a unique number, the IMEI. This targets particularly the theft of mobile equipment.

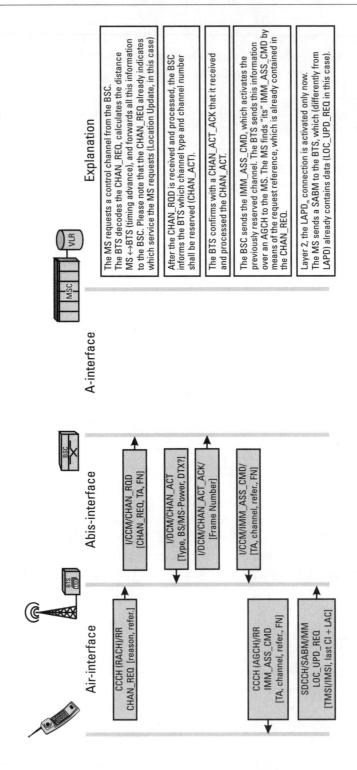

Figure 12.2 Location update on the BSS interfaces.

Scenarios

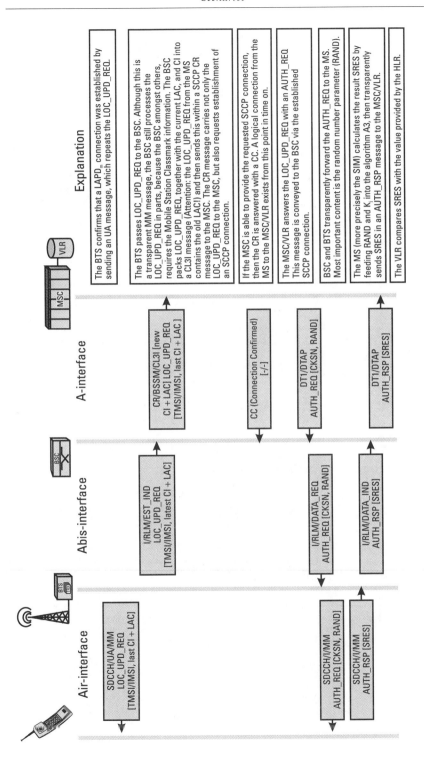

Figure 12.2 (continued)

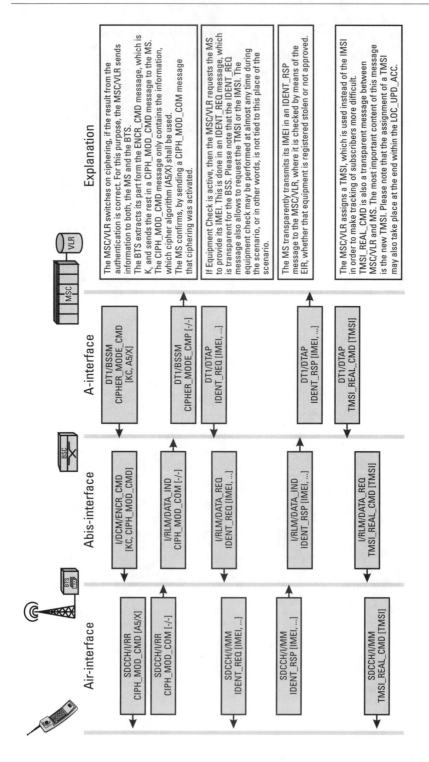

Figure 12.2 (continued)

Scenarios

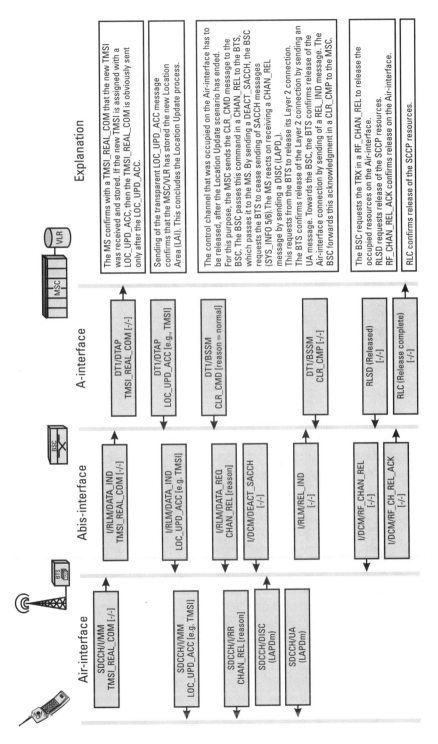

Figure 12.2 (continued)

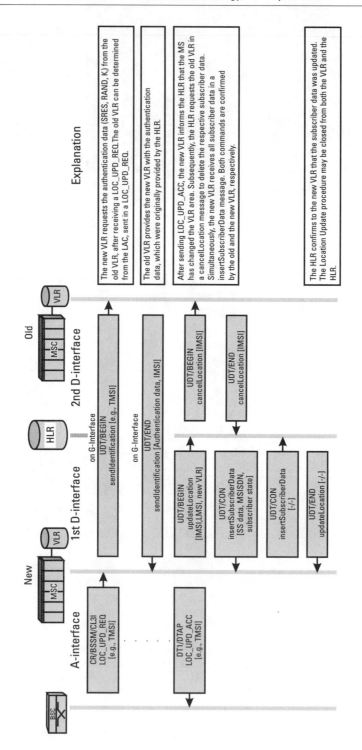

Figure 12.3 Location update on the NSS interfaces.

For that purpose, the NSS includes a database, the EIR, that contains information such as the serial numbers of barred mobile equipment. Figure 12.4 illustrates the process of an equipment check between MSC/VLR and EIR.

12.3 Mobile Originating Call

12.3.1 Mobile Originating Call in the BSS

The MS initiates a network access with a MOC. The network access is specified in more detail in the first message that the MS sends (CM_SERV_REQ) on the SDCCH. The reasons for such a request could be a regular telephone call, transfer of MO-SMS, activation of a supplementary service, or an emergency call. The CM_SERV_ACC message, shown in Figure 12.5 as a confirmation of a CM_SERV_REQ, is used only when ciphering is not active. If ciphering is active, the MS, when receiving the CIPH_MOD_CMD message, interprets it as a positive acknowledgment from the network for the service request. Figure 12.5 illustrates the MOC on the BSS interfaces. In addition, Figure 12.6 explains which signaling is taking place during an active connection.

12.3.2 Mobile Originating Call in the NSS

From the perspective of the NSS, a connection request of a subscriber can be directed as follows:

1. To the same PLMN (MS-to-MS call);
2. To another PLMN (MS-to-MS call);
3. To the ISDN (digital);
4. To the PSTN (analog).

In case 1, ISUP signaling is used between both MSCs, after the HLR of the called subscriber has provided the necessary routing information. ISUP is also used in cases 2 and 3. There is no principle difference between a call to another PLMN and to an ISDN.

Figure 12.7 shows signaling for the MOC, which applies to cases 1, 2, and 3. (Query of the routing information is discussed in Section 12.4.2.) In case 4 (PSTN), the gateway MSC has to provide the necessary conversion between digital and analog signaling.

234 GSM Networks: Protocols, Terminology, and Implementation

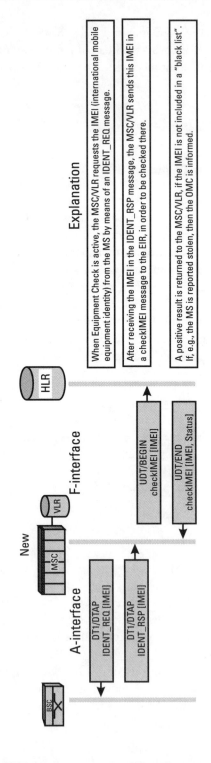

Figure 12.4 Scenario of checking the IMEI.

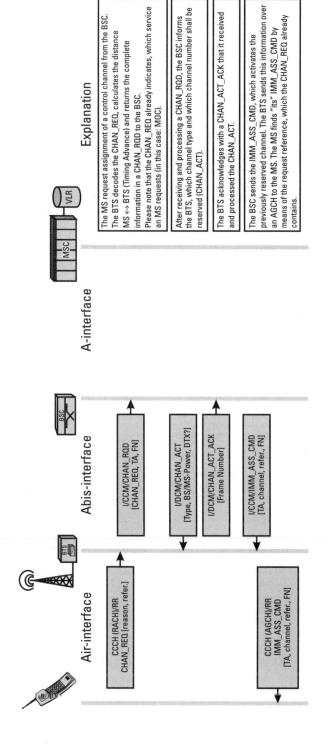

Figure 12.5 Mobile originating call in the BSS.

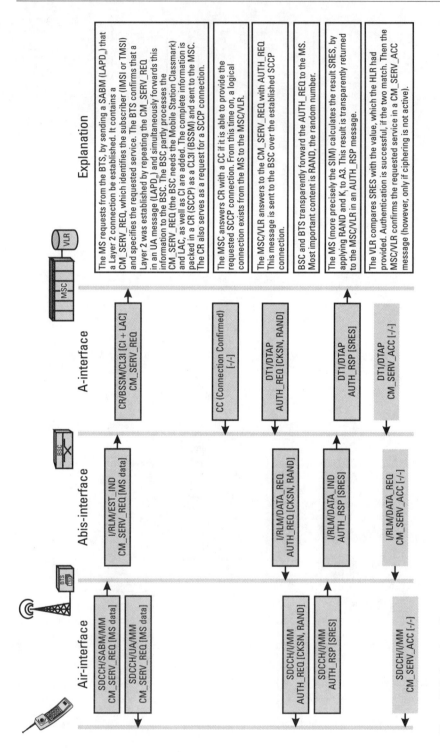

Figure 12.5 (continued)

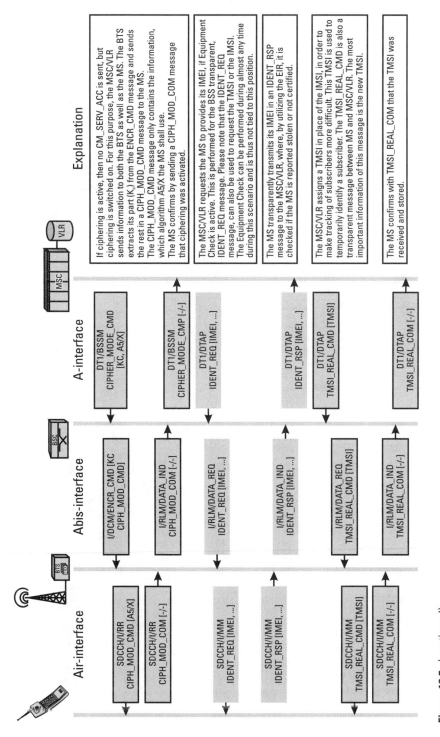

Figure 12.5 (continued)

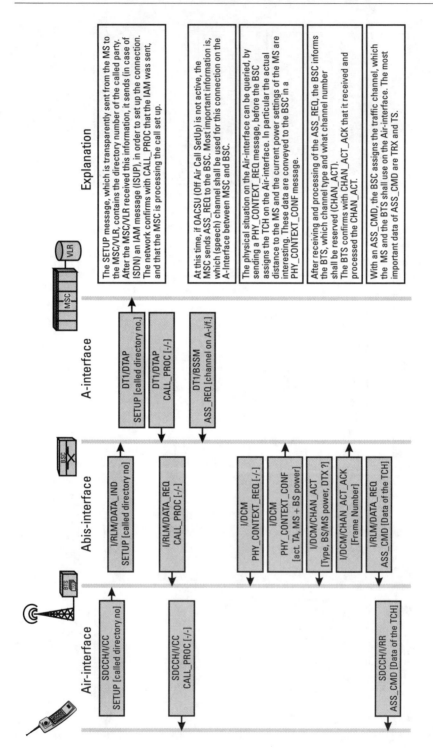

Figure 12.5 (continued)

Scenarios

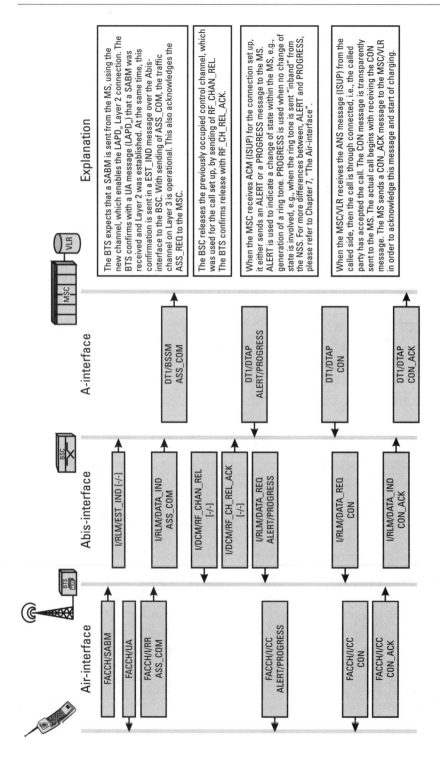

Figure 12.5 (continued)

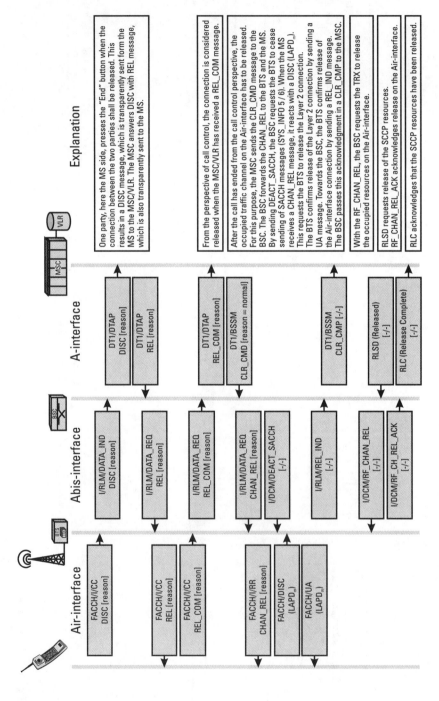

Figure 12.5 (continued)

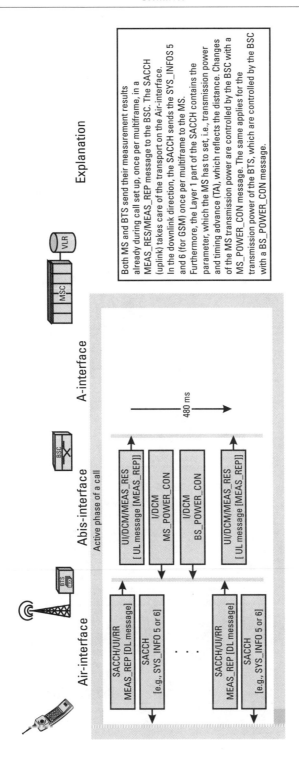

Figure 12.6 Signaling traffic during a connection.

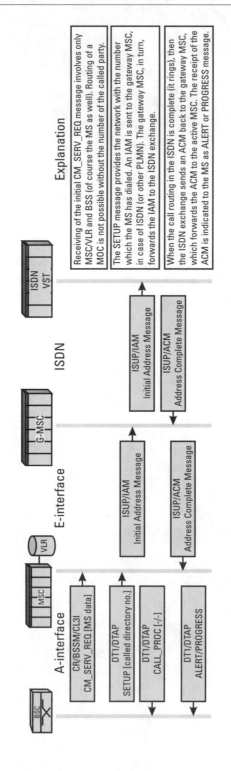

Figure 12.7 Mobile originating call in the NSS.

Scenarios 243

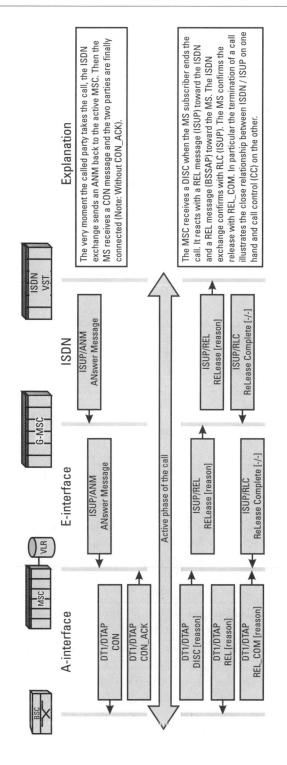

Figure 12.7 (continued)

12.4 Mobile Terminating Call

12.4.1 Mobile Terminating Call in the BSS

In the case of a MTC, a subscriber (PLMN internal or from external) tries to reach a mobile subscriber. Although the connection request in a MTC is not originated by the MS, there are some similarities between the MOC and the MTC, particularly for the network access on the BSS interfaces. The following differences, however, need to be pointed out:

- The CHAN_REQ of the MS in an MTC is the answer to PAGING_REQ.
- The first message sent over the new SDCCH is not a CM_SERV_REQ but a PAG_RSP.
- The SETUP message in an MTC is initiated by the MSC/VLR, not by the MS.

In the example of a MTC shown in Figure 12.8, the call is terminated by the "other" side. Thus, this example is the exact opposite of the example in Section 12.3. Of course, in real life, either side may end the call.

12.4.2 Mobile Terminating Call in the NSS

From the perspective of the NSS, finding a mobile subscriber is one of the most important tasks during an MTC. What is the scenario for this search? How is the further cooperation with BSS and ISDN performed? This section focuses on answering those questions.

In the case of a MTC, any subscriber from within or outside the PLMN dials the MSISDN of a subscriber. The ISDN routes the call to the PLMN or, more precisely, to the gateway MSC, based on the information contained in the MSISDN, national destination code (NDC), and the country code. This step is not necessary in case of a PLMN internal MS-to-MS call. After reception by the gateway MSC the HLR of the subscriber has to be identified, based on the MSISDN, to retrieve information, in particular the VLR area where the subscriber currently roams (the HLR does not know the location area). The HLR, in turn, queries the VLR, which assigns an MSRN for routing purposes and provides that number to the HLR. The HLR only forwards the MSRN to the gateway MSC, which uses the address to finally route the call to the destination MSC/VLR. After a radio connection to the MS is established, the NSS or, more precisely, the MSC mainly provides interfacing functionality between the

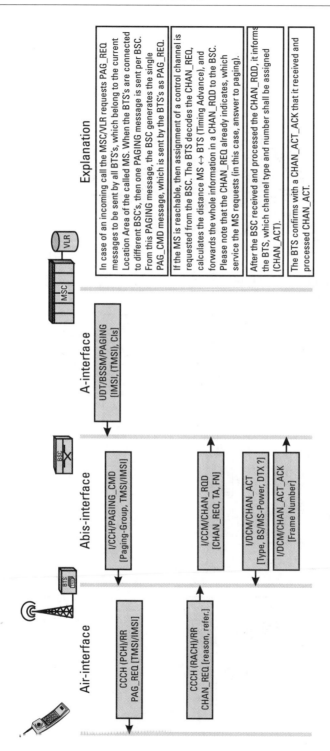

Figure 12.8 Mobile terminating call in the BSS.

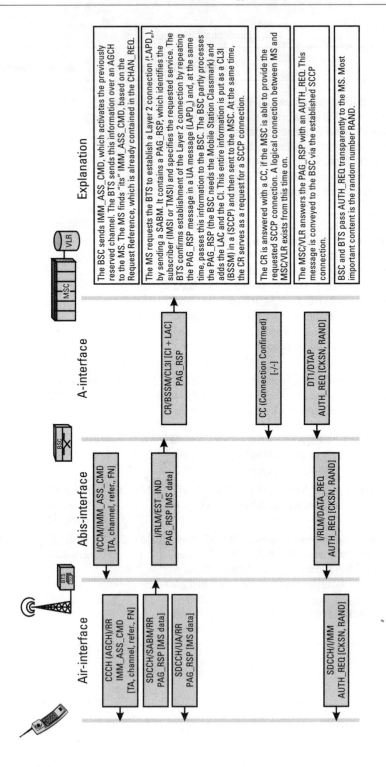

Figure 12.8 (continued)

Scenarios

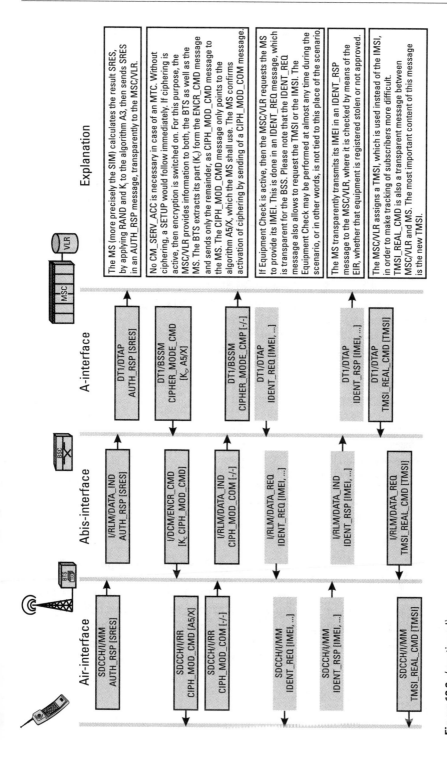

Figure 12.8 (continued)

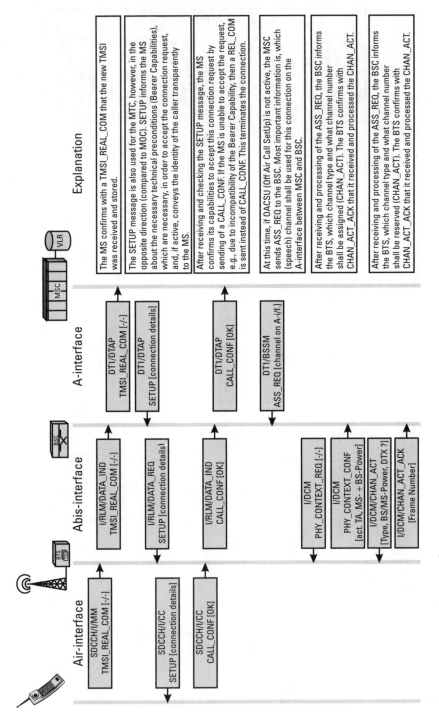

Figure 12.8 (continued)

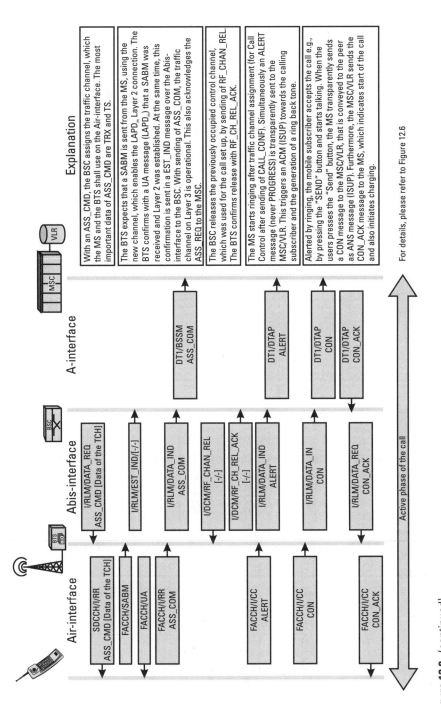

Figure 12.8 (continued)

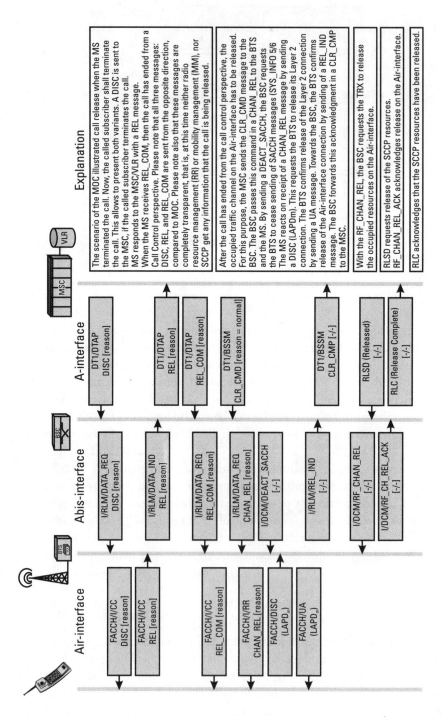

Figure 12.8 (continued)

NSS and the ISDN. For a call from the analog PSTN, the gateway MSC has to convert the analog signaling into digital SS7 ISUP signaling. (The reverse applies similarly, of course, in the opposite direction.) The process is illustrated in Figure 12.9.

12.5 Handover

The capability of a handover and its execution are some of the more interesting characteristics of a modern wireless service. Note that the GSM-specific term *handover* corresponds to the same concept as *handoff*, which is used in cellular networks in the United States. A handover is defined as the change of the currently used radio channel (SDCCH or TCH) to another radio channel during an existing and active connection between MS and BTS. The requirement for an already existing and active radio connection distinguishes the handover from the SDCCH assignment in the idle state. The differences relative to frequency redefinition, channel modify, and TCH assignment are, however, fuzzy. For the intra-BTS handover, it is even an option to use the TCH assignment procedure for handover. GSM has defined several handover types that can be activated by the network operator and executed when necessary. (The handover types are described in Section 12.5.3.) The decisions of whether a handover should be performed and if so the type of handover chosen lie with the BSC, based on the measurement results of the BTS and the MS (MEAS_RES/MEAS_REP), taking into account the parameter and criteria set by the OMC. A detailed explanation of these criteria and the possible parameters is beyond the scope of this presentation and provides enough material for another book. What is presented here are the starting points and values for a decision on handover by the BSC as the controlling instance, based on the measurement results sent to the BSC once per multiframe.

12.5.1 Measurement Results of BTS and MS

The example in Section 12.5.2 analyzes a MEAS_RES message in more detail. The reader is asked to duplicate the following values by means of that example. Table 12.1 provides an overview of the power control and most important parameters, which can be separated into two groups:

- The measurement results of the MS, which can be separated into the results related to the active BTS and those related to the neighbor cells;
- The measurement results of the BTS.

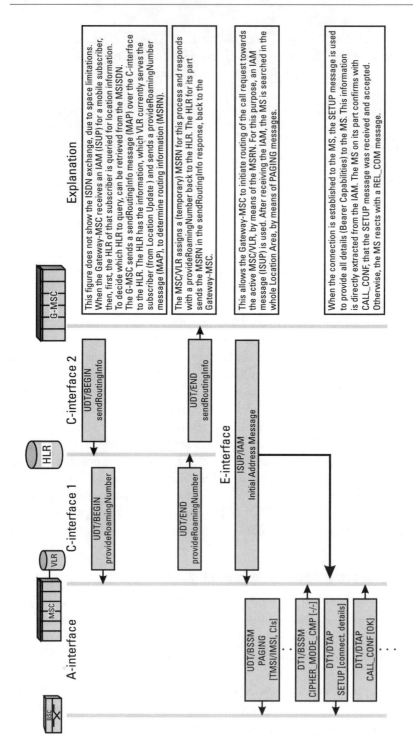

Figure 12.9 Mobile terminating call in the NSS.

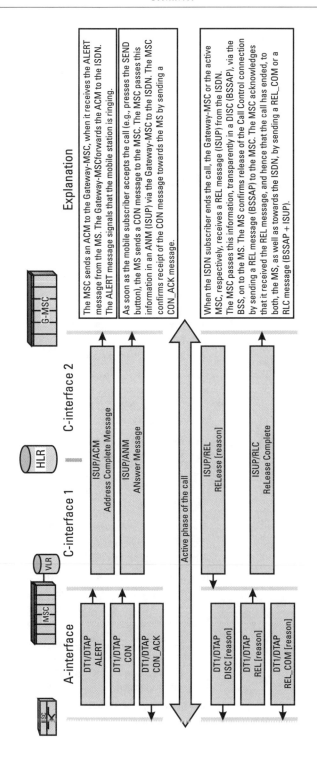

Figure 12.9 (continued)

The values for the receiving level (= RXLEV) and the receiving quality (= RXQUAL) are provided for both cases. Obviously, in the case of neighbor cell measurements, only the receiving level measurements are available, since the MS does not have an active connection.

Table 12.1
Input Parameters for Power Control and Handover Decision

Measured Value	Object to be Measured	Explanation
RXLEV-FULL-SERVING-CELL (6 bit $0 \equiv 63_{dec}$)	BTS	With what level does the MS receive the active BTS? This value is always provided but only relevant if no DTX is active in the downlink direction.
RXLEV-SUB-SERVING-CELL (6 bit $0 \equiv 63_{dec}$)	BTS	With what level does the MS receive the active BTS? This value is always provided but only relevant if DTX is active in the downlink direction.
RXQUAL-FULL-SERVING-CELL (3 bit $0 \equiv 7_{dec}$)	BTS	How well (BER) does the MS receive the active BTS? The BER is determined by means of the Training Sequence Code. This value is always present but only relevant if no DTX is active in the downlink direction.
RXQUAL-SUB-SERVING-CELL (3 bit $0 \equiv 7_{dec}$)	BTS	How well (BER) does the MS receive the active BTS? The BER is determined by means of the Training Sequence Code. This value is always present but only relevant if DTX is active in the downlink direction
RXLEV-NCELL 1 - N (6 bit $0 \equiv 63_{dec}$)	BTS	With what level does the MS receive the neighbor cells (as indicated in SYS_INFO 2)?
RXLEV-FULL-up (6 bit $0 \equiv 63_{dec}$)	MS	With what level does the BTS receive the MS? This value is always provided but only relevant if no DTX is active in the uplink direction.
RXLEV-SUB-up (6 bit $0 \equiv 63_{dec}$)	MS	With what level does the BTS receive the MS? This value is always provided but only relevant if DTX is active in the uplink direction.
RXQUAL-FULL-up (3 bit $0 \equiv 7dec$)	MS	How well (BER) does the BTS receive the MS? The BER is determined by means of the Training Sequence Code. This value is always present but only relevant if no DTX is active in the uplink direction.
RXQUAL-SUB-up (3 bit $0 \equiv 7dec$)	MS	How well (BER) does the BTS receive the MS? The BER is determined by means of the Training Sequence Code. This value is always present but only relevant if DTX is active in the uplink direction.
Timing Advance (6 bit $0 \equiv 63dec$)	MS	How far is the MS away from the BTS?

12.5.2 Analysis of a MEAS_RES/MEAS_REP

This example (Figure 12.10) analyzes a MEAS_RES in more detail (including a MEAS_REP) that was captured with a protocol tester on the Abis-interface of a GSM900 PLMN. Note, when looking at the values for RXQUAL or RXLEV, respectively, that the following applies: The qualifier "FULL" or "ALL" refers to the case with *inactive* DTX, while the qualifier "SUBSET" refers to the case with *active* DTX.255

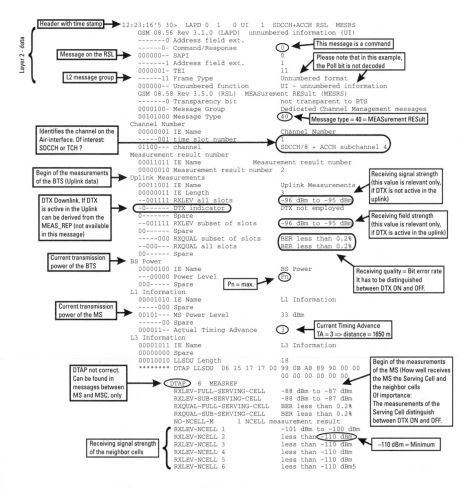

Figure 12.10 Analysis of a MEAS_RES message.

Whether DTX is active in the *uplink* direction can be derived from the *downlink* measurements (MEAS_REP). Whether DTX is active in the *downlink* direction can be derived from the *uplink* measurements (MEAS_RES).

When tracing a call protocol, one occasionally sees MEAS_RES messages, which do not contain a MEAS_REP. That is the case if the BTS is unable to decode the uplink SACCH, because of problems in the area of radio transmission. In that case, typically, the connection breaks down shortly due to radio link timeout. This type of call drop cannot be prevented completely. A closer investigation is necessary, however, if such failure rate is relatively high or is increasing dramatically.

12.5.3 Handover Scenarios

When categorizing the various handover scenarios, two criteria or perspectives have to be considered:

- What entity is executing the handover? In other words, which BTS is the source and which BTS is the destination of a handover? This category covers expressions like intra-BTS and inter-MSC. We will return to this issue, because the location of the executing functionality is important for the handover scenario. Note that the BSC always decides whether a handover needs to be executed. However, for the execution itself, either the BSC or the MSC might be in charge.
- Are the system clocks of the origination cell and the destination cell of a handover finely synchronized? This criterion affects the handover scenario on the Air-interface.

Regarding the second criterion: it has to be added that GSM, regarding synchronization, distinguishes among four different levels: *nonsynchronized*, *synchronized*, *presynchronized*, and *pseudo-synchronized*. Details and differences of these concepts are provided in the Glossary.

The protocol scenario is the same for synchronized, presynchronized, and pseudo-synchronized handover. Only the nonsynchronized handover has to be regarded separately.

12.5.3.1 Synchronized Versus Nonsynchronized Handover

In practical operation, synchronized and nonsynchronized handover are mainly used, where nonsynchronized handover has the advantage of shorter dead time during the handover process. Synchronized handover, on the other hand, requires that the clock of all involved cells be finely synchronized. Therefore, the synchronized handover can be used only in intra-BTS handover or inter-BTS handover when the BTSs are sectorized or collocated, in which case the equipment is located closely enough together to allow for fine synchronization in an inexpensive way. However, the majority of the handovers performed in commercial systems are nonsynchronized.

12.5.3.2 Intra-BTS Handover

In intra-BTS handover, a new channel in the same BTS is assigned to the MS. The intra-BTS handover does not distinguish whether the new channel is just on another timeslot in the same TRX (frequency) or whether the TRX changes as well. An intra-BTS handover is performed particularly when the RXQUAL values in uplink or downlink are relatively bad, while the RXLEV values stay good (interference). Figure 12.11 illustrates the intra-BTS handover. The procedure usually is executed autonomously by the BSC, but the MSC also may be in charge. It is worth pointing out that an intra-BTS handover is always synchronized, since all TRXs of a BTS have to use the same clock. Figure 12.12 presents the corresponding scenario. Note that for an intra-BTS handover both the ASS_CMD message and the HND_CMD message may be applied.

12.5.3.3 Intra-BSC Handover

In the intra-BSC handover, an MS changes the BTS but not the BSC, as illustrated in Figure 12.13. Like the intra-BTS handover, the intra-BSC handover may be carried out autonomously by the BSC, without support from the MSC. It is an option by the network operator, however, to decide that the MSC supervises the process. For intra-BSC handover, depending on the circumstances, both handovers are possible, synchronized as well as nonsynchronized. Figure 12.14 presents the scenario for a nonsynchronized intra-BSC handover. The difference from the synchronized intra-BSC handover is simply that the PHYS_INFO messages from the BTS would not have to be sent.

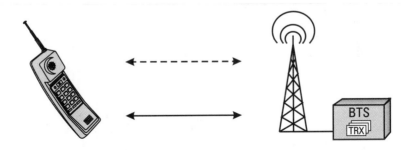

Figure 12.11 The intra-BTS handover.

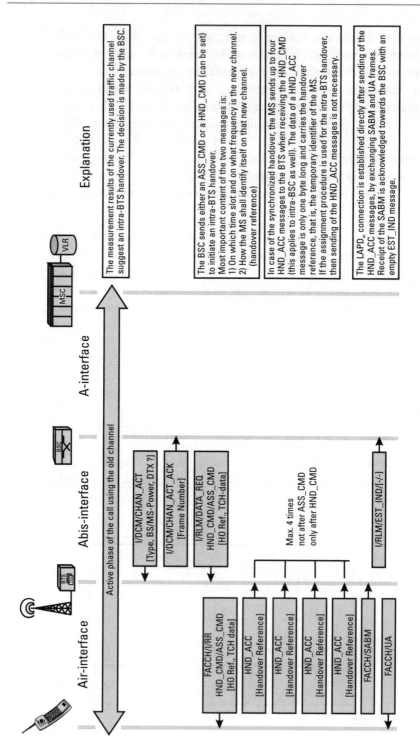

Figure 12.12 The synchronized intra-BTS handover.

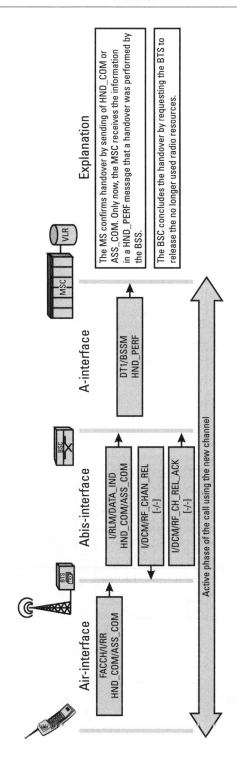

Figure 12.12 (continued)

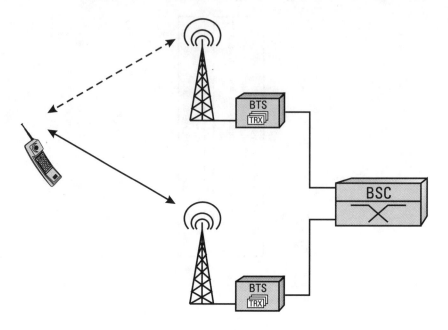

Figure 12.13 The intra-BSC handover.

12.5.3.4 Intra-MSC Handover

In an intra-MSC handover, the MS changes not only the BTS, but the BSC as well (Figure 12.15). Therefore, external handover is another term for this type of handover. In contrast to the intra-BTS handover and the intra-BSC handover, the MSC mandatorily is in charge for the execution of the intra-MSC handover. The responsibility for the MSC does not, however, include processing the measurements of the BTS or the MS or to conclude that a handover is necessary. These functions always remain with the BSC. (Note: How should the BSC know in advance what type of handover will be needed, or, in other words, where the destination cell of a handover will be?) After the BSC has decided that a handover to a BTS controlled by another BSC has to be performed, it informs the MSC and waits for instructions. The MSC requests the target or new BSC to reserve the necessary radio resources for the new connection. If the new BSC is able to process the request, it returns a complete HND_CMD message to the MSC, which will pass it on to the first BSC. The old BSC, after receiving the response from the target BSC, passes the HND_CMD to the MS and releases the radio resources after confirmation by the MSC. Figure 12.16 describes this scenario. In Figure 12.16, the distinction between messages to and from the old and new BSC is made by the different arrow styles (old BSC = small arrows, new BSC = large arrows).

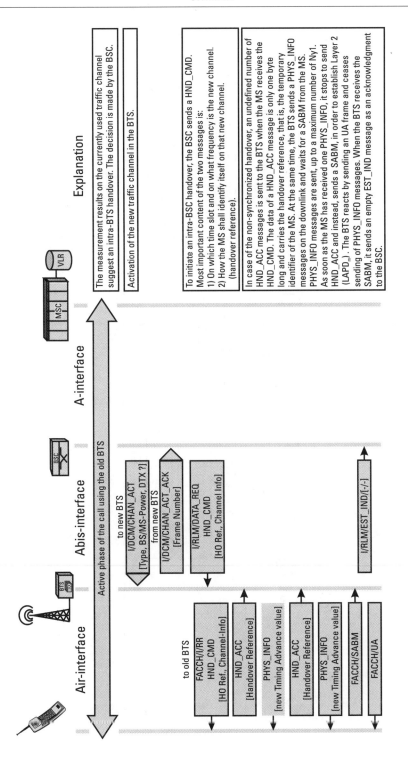

Figure 12.14 The nonsynchronized intra-BSC handover.

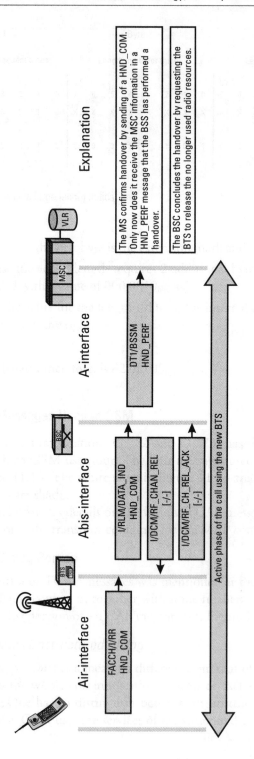

Figure 12.14 (continued)

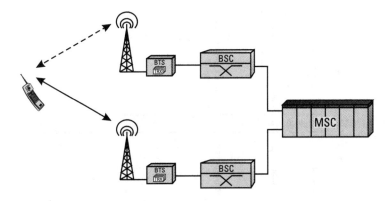

Figure 12.15 The intra-MSC handover.

12.5.3.5 Inter-MSC Handover and Subsequent Handover

Inter-MSC Handover

The handover cases described so far have involved increasingly more and more subsystems that were affected every time the next higher level in the hierarchy took over control of the execution of the handover process. However, which part of the system will take the control task when even the MSC is changed during a handover? This scenario is referred to as inter-MSC handover.

The answer is simple but requires some explanation. The control of the inter-MSC handover stays with the old MSC (MSC A in Figure 12.17).

Furthermore, even the CC functions for a connection handed over to MSC B stay with MSC A *after* the inter-MSC handover, that is, MSC A stays in full control of that connection.

Consider this example. If a GSM call toward an ISDN subscriber is originally set up in city A (with MSC A), that call is controlled by MSC A in city A. If the mobile subscriber drives into another MSC area, let's say city B (with MSC B), the call needs to be handed over by an inter-MSC handover procedure from MSC A to MSC B. However, because the ISDN does not provide any handover capabilities, the outgoing connection stays somewhat tied to MSC A. Therefore, MSC A remains in charge for all CC functions and, by means of the inter-MSC handover, just relays its RR functions toward MSC B. This rule is not restricted to calls toward the ISDN but is applicable for all calls. Figure 12.18 shows the scenario for an inter-MSC handover.

If in an inter-MSC handover, the original MSC always maintains the CC functionality, what then are the tasks of MSC B, and what happens if another inter-MSC handover is performed, either back to MSC A or to a third MSC, MSC C? As before, these answers are simple but require some explanation.

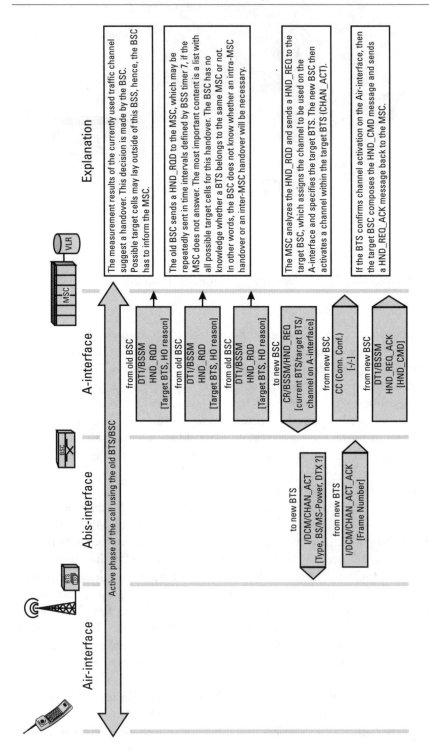

Figure 12.16 Scenario of intra-MSC handover.

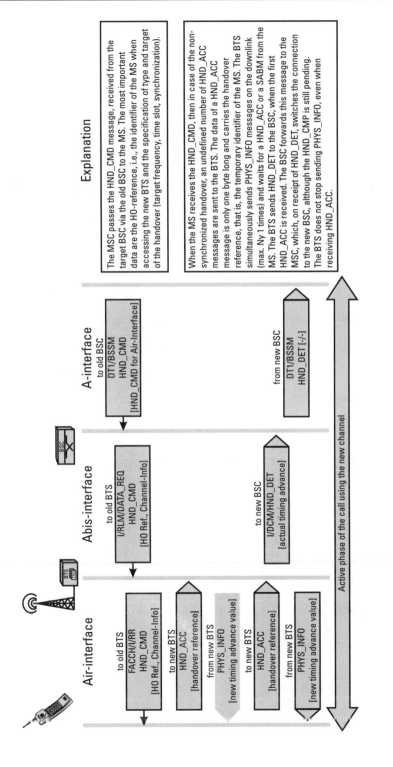

Figure 12.16 (continued)

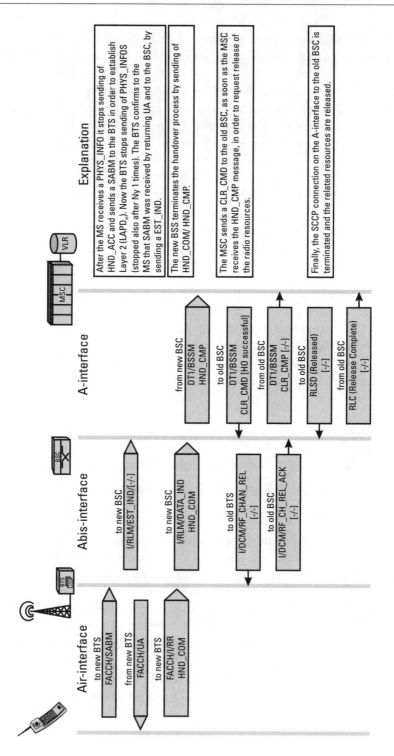

Figure 12.16 (continued)

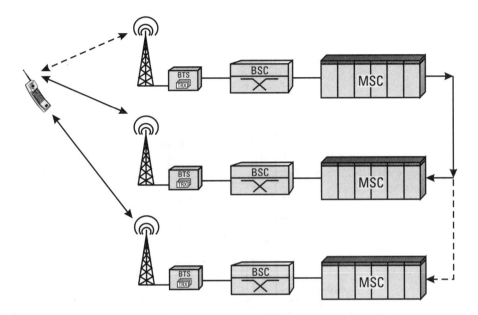

Figure 12.17 Inter-MSC and subsequent handover.

As already mentioned, MSC B, as the receiving MSC, takes control over all RR tasks, which are internal to the MSC. These are in particular intra-MSC handover-related tasks.

The subsequent handover scenario is applied in case of such a second inter-MSC handover, described next.

Subsequent Handover

Two cases have to be distinguished for the subsequent handover:

- A handover from MSC B back to the original MSC (MSC A);
- A handover into a third MSC area (MSC C).

Although MSC B initiates a subsequent handover, MSC A still maintains overall control of the call, during and even after a subsequent handover. Overall control means that, in any case, independent of whether the target of the handover is MSC A or a third MSC, the CC functionality always remains with MSC A. Figure 12.19 provides the scenario and the messages for a subsequent handover back to MSC A. Figure 12.20 provides the scenario and the messages for a subsequent handover to a third MSC (MSC C).

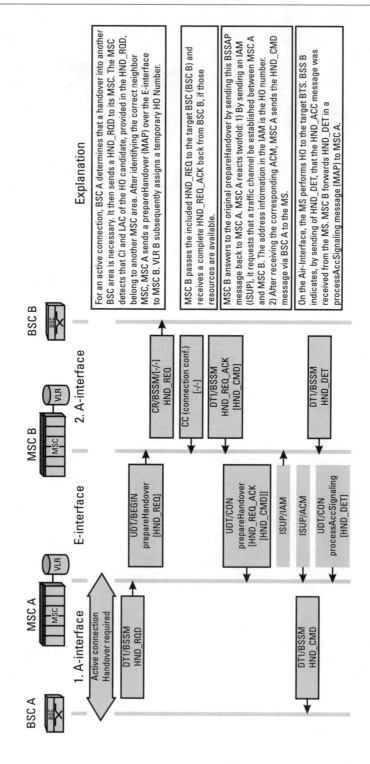

Figure 12.18 Scenario for inter-MSC handover.

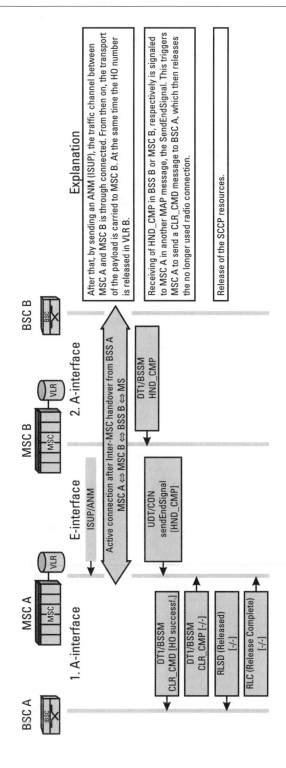

Figure 12.18 (continued)

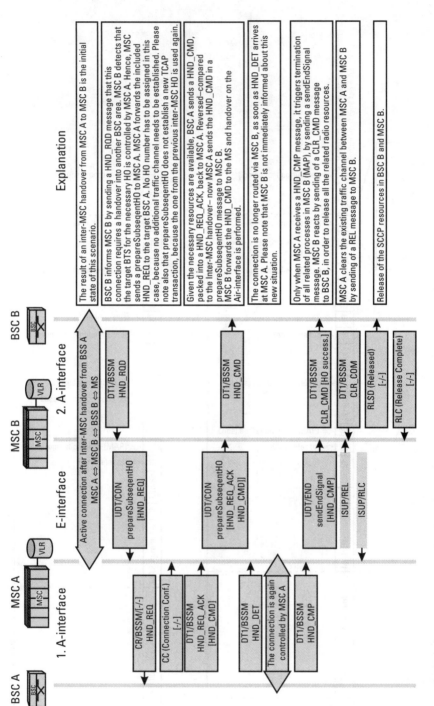

Figure 12.19 Senario for subsequent handover back to MSC A.

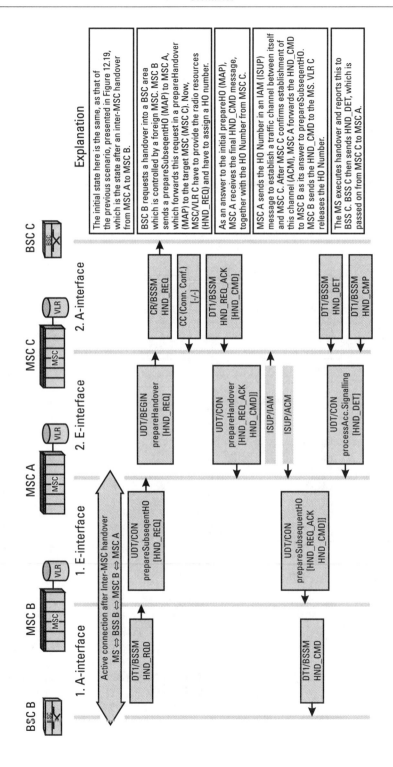

Figure 12.20 Senario for subsequent handover to MSC C.

272 GSM Networks: Protocols, Terminology, and Implementation

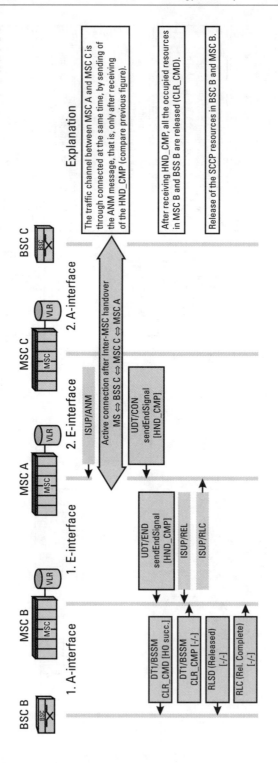

Figure 12.20 (continued).

Message Transport Over the E-Interface

MSCs have to relay BSSAP-related data that is coming from the A-interface, over the E-interface during inter-MSC handover, during subsequent handover, and thereafter. For that reason, the MAP protocol includes two messages specifically defined to perform that task, that is, to transparently transmit messages between MSCs. The two messages are the forwardAccessSignaling message and the processAccessSignaling message. As illustrated in Figure 12.21, the forwardAccessSignaling message is used to transport BSSAP messages in the downlink, that is, to the MS, while the processAccessSignaling message is used in the opposite direction.

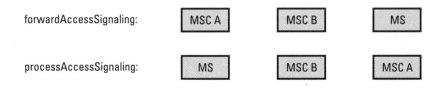

Figure 12.21 Transfer of BSSAP signaling over the E-Interface after Inter-MSC handover.

13

Quality of Service

The term quality of service (QOS) has a specific meaning in telecommunications. It refers to the usability and reliability of a network and its services. This chapter highlights possible applications of protocol analyzers for the supervision and analysis of network quality and presents the problems that GSM network operators most frequently face.

How can protocol test equipment be used to improve network quality? What errors occur most frequently in GSM and what are the root causes of those errors? How can reliable and comparable routines for statistical analysis be developed? This chapter focuses on answering those questions.

13.1 Tools for Protocol Measurements

In general, there are three ways to determine the QOS of a GSM network:

- Drive tests. Teams, paid by the network operator, drive on predefined routes in the network and periodically initiate calls. The results (e.g., unsuccessful handover, low-quality audio, dropped calls, signal strength) are transferred from the MSs to a dedicated PC, where the respective data are available for postprocessing. This kind of measurement represents most closely the network quality as real subscribers experience it. The disadvantage, however, is that only a limited area and a small time window can be tested and the testing is extremely expensive. Real-time measurement tools for call quality like QVOICE

belong to the same category but allow for a more objective judgment on the call quality.
- Protocol analyzers. Preferably at a central place, protocol analyzers are connected to BTSs, BSCs, and MSCs over a period of time. Frequently, these devices come with remote access capabilities, which eases most configuration changes. Captured trace files can be uploaded to a central office for statistic evaluation. When problems are detected, the trace file needs to be analyzed manually and more thoroughly. Measurements with protocol analyzers have the advantage that all captured events are available for later, detailed analysis. The disadvantage is that it is not commercially feasible to have them in such great numbers that an entire GSM network could be observed permanently. (Of course, vendors of protocol analyzers might have a different opinion.)
- OMC. Several counters can be activated in the BSS and in the NSS from the OMC to provide the central office with the most important data about network quality. Various counters for all kinds of events permanently provide the network operator with information about the state and quality of the network. Examples are counters for the number of incoming or outgoing handovers; call drops before, during, and after assignment; and dropped calls due to missing network or radio resources. The major advantage of measurements via the OMC is that it provides results about the quality of the entire network rather than single BTSs or BSCs. On the other hand, this method is the most abstract one and the one that least relates to what the subscribers are encountering. Furthermore, it hardly allows for a detailed postmortem analysis.

Only the combination of the measurement results of OMC, protocol test equipment, and the–most expensive–drive tests, allows to make a qualified and objective statement about the real Quality of Service.

The focus of the following explanation is put on measurements with protocol analyzers, whereas this information can be translated, fairly easily to OMC measurements.

13.1.1 OMC Versus Protocol Analyzers

QOS involves the permanent observation, supervision, and adjustment of the various network parameters in a telecommunication system. GSM makes no exception. One of the most important functions of the OMC is to provide feedback about network quality in real-time (more or less). For that purpose, a network operator can choose from a large selection of measurable events that may occur in the BSS or NSS, to be gathered in a predefined time period and then displayed at the OMC. It is a typical task of the network operator to determine the network quality from the available data and to decide on eventual corrective measures.

The time period that a single TS is occupied, the rate of ASS_FAI, separated for MOC and MTC, and the number of handovers, based on DL_QUALITY, are more examples of such counters, to name only a few.

The OMC has some advantages in detecting existing and potential network problems, compared to protocol analyzers:

- The OMC is part of the standard delivery of a GSM system and as such requires no extra procurement.
- The OMC plays a central part in its function as an O&M platform. If network quality is monitored by the OMC, countermeasures can be taken immediately when problems arise.
- Although protocol analyzers allow the capture of data on the A-interface, other important data have to be gathered on the BTS level, that is, on the Abis-interface. A complete analysis of data related to the entire coverage area by protocol analyzers, including data from the Abis-interface (e.g., idle channel measurements on the Air-interface) is not possible, due to logistical and financial limitations. That is where the OMC appears as the optimal tool.
- The OMC measures events and provides the results of respective counters to the network operator. Those values later can be processed and evaluated. For that reason, the OMC is the optimal aid for statistics tasks.

The protocol analyzers, on the other hand, can also claim a number of advantages as follows:

- A protocol analyzer is not part of the standard delivery of the infrastructure but an *independent* tool. To analyze the OMC counters, the field engineer needs system-specific know-how for the measurement. For the protocol analyzer system-independent know-how is necessary.
- The OMC is of little help for problem analysis during the process of bringing a network or single network elements into service (without subscriber traffic). That is different for protocol analyzers.

- Protocol analyzers allow the network operator and the system supplier to conduct measurements on the lowest level; it allows capture of complete call traces.
- Protocol analyzers typically come with additional software that provides for excellent statistical analysis. That enables the field engineer to evaluate a situation quickly. Technologically advanced software makes a manual analysis of data unnecessary in most cases.

In summary, both OMC and protocol analyzers are necessary tools for the network operator as well as for the system supplier. That is particularly true for the supervision of QOS. OMC and protocol analyzers perfectly supplement each other in this area, to ensure optimal network quality for both the network operator and subscribers. For those reasons, network operators use the OMC to monitor a GSM network on a broad scale, while they still have specialists with protocol analyzers for the bottom-down analysis of specific network problems. If the OMC detects a problem but is not able to identify the cause of the problem, the specialist goes on site, usually with a protocol analyzer, to narrow the problem down.

In the laboratory of a system provider or for product development and integration, the protocol analyzer plays a much more important role than the OMC.

13.1.2 Protocol Analyzer

A protocol analyzer is a measurement tool with a high impedance that can be connected between two system devices to intercept the digital data traffic between the devices (Figure 13.1). The focus here is on the analysis of signaling, that is, the control information between the devices. For that purpose and to simplify its operation, a state-of-the-art protocol analyzer needs the following:

- The complete hardware and software necessary to detect and adapt to the settings of the various interface configurations and time slots;
- Software that allows decoding of the binary PCM-code in plain text;
- A screen to display the measurement results;
- Facilities for remote access;
- Storage capacity of an appropriate size to store the measurement results for postprocessing;

Quality of Service

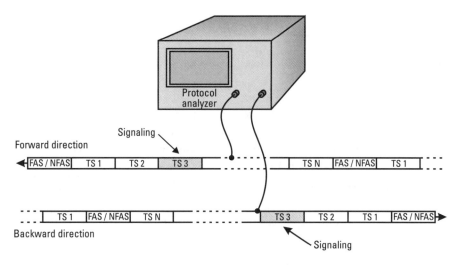

Figure 13.1 Setup of a protocol analyzer.

- Software tools to filter or search for specific data and parameters;
- A software interface to regular PC editor to be able to export data, for example, for reports;
- Flexible and easy-to-use statistic functionality.

The majority of the protocol analyzers available today meet those requirements reasonably well. Operation is fairly simple, because most of the devices are PC-based and offer additional hardware in form of expansion boards.

The protocol analyzer of the late 1990s consists of a mechanically robust PC expansion board loaded with software and offers the above listed functionality and additional features. Built into a laptop computer or desktop PC, protocol analyzers are portable or can be used in versatile ways in the laboratory.

It is worthwhile to point out what enormous progress has been made in recent years by the manufacturers of such equipment. While there was hardly any choice in the beginning of GSM (1991), system manufacturers and network operators quickly reacted to the increased need. Today, a multiplicity of protocol analyzers are available from many European and U.S. vendors. With GSM, a new era has started in this area. High-frequency (radio) measurements clearly have lost importance, compared to protocol measurements. GSM, as a digital standard, carries the analog problems on the Air-interface (e.g., interference), digitally coded into the BSS. Most radio-specific problems are well suited for analyzing and narrowing down the problem with a protocol analyzer.

13.2 Signaling Analysis in GSM

Most of the signaling measurements in GSM are performed on the interfaces in the BSS for two reasons:

- The majority of the subsystems and interfaces of a wireless network are part of the BSS. Simply from the sheer number of Abis-interfaces and A-interfaces, BSCs, and BTSs, it is obvious that more need for error analysis exists in this area.
- The critical area of a wireless system, however, is the Air-interface. Interference and handover problems are only two examples for mobile-specific error scenarios that can best be investigated in the BSS.

Measurement of signaling is, of course, also important in the NSS. Note that there is an interesting difference between measurements in the BSS and the NSS:

Measurements of signaling in the NSS more often are performed manually, that is, individual messages have to be analyzed. The reason is that the problems in the NSS typically are not hardware related but are concerned with interworking of protocol implementations of different manufacturers or related to software errors after major or minor software updates.

The majority of the problems that occur in the BSS, on the other hand, are hardware related or caused by deficiencies in network tuning. These types of errors typically can be detected only after a statistical, that is, an automatic, analysis.

13.2.1 Automatic Analysis of Protocol Traces

In most cases, it is important to localize the source of error or first to get an overview. For this application, a number of manufacturers have developed extensive software packages that illustrate a summary or interpretation of the most important results of such measurements graphically or in tabular form.

One example is the software package developed by a group of people from DeTeMobil, headed by Mr. Hochscherff. Their software package illustrates in tabular form error rates, quality data, and load parameter for various GSM interfaces. For the time being, it allows processing of measurement results captured with a SIEMENS K1103 or Wandel & Goltermann MA-10. Figures 13.2a, 13.2b, and 13.3 show parts of such analysis on the A-interface and the Abis-interface, which are presented here with the permission of DeTeMobil. Note the extensive amount of detailed information that the

```
******************** A-Interface Statistic ***-1.9.6d-**********************
Copyright by M.Dicks, W.Ditzer /DeTeMobil GmbH
Starttime: 06/06/1997 10:07              Stoptime : 06/06/1997 21:18
MSC: Entenhausen(A)  1.BSC spc: 12345 Own-BTSs: 33 LAs: 2 TotalBTSs: 43
Z  0   1.0 MOC-Analysis (no EC/SS/SMS)     Incoming
Z  1   Cm Serv Req       :   26543  100.0% HO Request        :       0  100.0%
Z  2   Cm Serv Acc       :      49    0.2% HO Request Ack    :       0    0.0%
Z  3   Cm Serv Rej       :     143    0.5% HO Complete       :       0    0.0%
Z  4   Setup             :   25539   96.2%
Z  5   Alerting          :   21183   79.8% 6.0 Other messages
Z  6   Connect           :   14875   56.0% Auth Request      :   87192  100.0%
Z  7   Connect Ack       :   14778   55.7% Auth Response     :   85715   98.3%
Z  8                                       Auth Reject       :       2    0.0%
Z  9   2.0 MTC-Analysis                    Cipher Cmd        :   84712
Z 10   Pagings           :   60415         Cipher Complete   :   84328
Z 11   Paging Response   :   13615  100.0% Ident Request     :    4388
Z 12   Setup             :   13125   96.4% Ident Response    :    4280
Z 13   Alerting          :   12721   93.4% TMSI Command      :   39704
Z 14   Connect           :    8305   61.0% TMSI Complete     :   83969
Z 15   Connect Ack       :    8266   60.7% MM-Status         :      20
Z 16                                       Block             :      32
Z 17   3.0 Location Update-Analysis        Block Ack         :      32
Z 18   LU Request        :   46598  100.0% UnBlock           :      30
Z 19   davon Normal      :   32802         UnBlock Ack       :      35
Z 20         Periodic    :    4139         Reset             :       0
Z 21         IMSI att    :    9657         Reset Ack         :       0
Z 22   LU Reject         :    1182    2.5% Reset CIC         :     169
Z 23   LU Accept         :   44733   96.0% Overload          :       0
Z 24                                       HoRqd SDCCH-OWN   :     252
Z 25   4.0 Assignment Analysis             HO Failure        :    1026
Z 26   Assignment Cmd    :   37809  100.0% HO Performed      :     160
Z 27   Assignment Cmp    :   37173   98.3% Clear Command     :  115487  100.0%
Z 28   Assignment Fail   :     612    1.6% Clear Complete    :  115481  100.0%
Z 29                                       Clear Request     :    3571    3.1%
Z 30   5.0 HO Analysis                     Pagings           :   60415
Z 31   5.1 Intra-BSC                       Paging Repeat     :   26039
Z 32   HO Required(TCH)  :   18876  100.0% 2.Pag & No Resp   :   26045
Z 33   HO Request        :   18814   99.7% SCCP-UDT          :   60882
Z 34   HO Request Ack    :   18753   99.3% SCCP-CREQ         :  115701
Z 35   HO Command        :   18445   97.7% SCCP-CC           :  115694
Z 36   HO Complete       :   17554   93.0% SCCP-DT1          : 1241708
Z 37   5.2 Inter-BSC                       SCCP-RLSD         :  115686
Z 38   Outgoing                            SCCP-RLC          :  115570
Z 39   HO Required(TCH)  :    2463  100.0% SCCP-CREF         :       1
Z 40   HO Command        :    2406   97.7% SCCP-IT           :     774
Z 41   HO Success        :    2299   93.3% SCCP-lost(BSC)    :      17
Z 42   Incoming                            SCCP-lost(MSC)    :      55
Z 43   HO Request        :    2595  100.0% MTP-TFA           :       0
Z 44   .....
Z 45   ...
Z 46   .
Z 47   .
```

Figure 13.2(a) Result of a measurement on the A-interface (BSC related).

```
Copyright  M. Dicks, W. Ditzer / DeTeMobil GmbH

Z 76    Cell Individuell Statistics Subpage:   1

Z 77  BSC Index            :       1       1       1       1       1       1       1       1       1       1
Z 78  Cell Id              :     123     234     345     456     567     678     789     890     901    4321
Z 79  LUs                  :    7276     756    1635    2250    1915    1165    1386    2102    1619    1519
Z 80  MOCs                 :    2763     251    1377     827    2642    1006    1768    1846    1662    2721
Z 81  MTCs                 :    1416     114     560     450    1389     470    1047     905     802    1321
Z 82  AssReq               :    3864     340    1821    1192    3703    1333    2608    2572    2324    3748
Z 83  AssCmpl              :   97.3%   98.2%   98.5%   98.2%   98.6%   98.3%   99.3%   98.8%   99.4%   97.8%
Z 84  AsFailCause 10 :        33.7%   33.3%   53.8%   57.9%   36.5%   64.7%   37.5%   68.6%   57.1%   27.7%
Z 84  AsFailCause  0 :        22.4%   66.7%   46.2%   31.6%   40.4%   35.3%   56.2%   25.7%   28.6%   18.1%
Z 84  AsFailCause 33 :        41.8%                                    23.1%            6.2%    5.7%    7.1%   53.0%
Z 84  AsFailCaus.oth.:        3.1%                   25.0%                                                          16.7%    1.7%
Z 85  .....
Z 86  ...
Z 87  ..
Z 88  .
```

Figure 13.2(b) Result of a measurement on the A-interface (BTS related).

```
             ------------ Analysis TCHCHECK 1.3 (DTC PDM F14) ------------
File: Entenh_1.rec
Begin of recording 03.09.1993 at 17:17h,
Duration 18.7 hours
Link 1-2; Channels:  |  33  |  92  |
Assignments          | 176  |  86  |

1:    act| assg| acpl| ho| hcpl| cf| 02| 1f| 28| ei|  Tall | Dlev| Ulev| Dqul| Uqul| SACCH%
        0|    0|   0%|  0|   0%|  0|  0|  0|  0|  0|    0  |   0 |   0 |  0 0|  0 0|  0.00
        0|    0|   0%|  0|   0%|  0|  0|  0|  0|  1|    0  |   0 |   0 |  0 0|  0 0|  0.00
       35|   15| 100%| 20| 100%|  0|  0|  0|  0|  0|  18.27 |  34 |  21 |  0 2|  0 2|  2.60
       70|   23|  96%| 47|  83%|  6|  4|  1|  0|  0|  44.58 |  36 |  22 |  0 2|  0 2|  2.40
       69|   37| 100%| 32|  72%|  2|  1|  1|  0|  0|  33.94 |  33 |  20 |  0 3|  0 2|  1.83
       70|   33| 100%| 37|  89%|  3|  2|  1|  0|  0|  38.89 |  31 |  20 |  0 2|  0 3|  2.33
       74|   36|  97%| 38|  89%| 16|  2|  5|  4|  0|  23.54 |  33 |  20 |  0 3|  1 6|  7.13
       74|   31| 100%| 43|  84%|  6|  2|  1|  2|  0|  32.22 |  35 |  22 |  0 2|  0 2|  3.77

2:    act| assg| acpl| ho| hcpl| cf| 02| 1f| 28| ei|  Tall | Dlev| Ulev| Dqul| Uqul| SACCH%
       26|   12|   0%| 14|   0%| 13|  5|  0|  8|  0|  0.05  |   3 |  18 |  7 49|  3 17|  45.45
       26|    9|   0%| 17|   0%| 12|  6|  0|  6|  0|  0.03  |   0 |  23 |  7 49|  4 26|  60.00
       27|   12|   0%| 15|   0%| 14|  6|  0|  8|  0|  0.02  |   0 |  25 |  7 49|  3 15|  72.73
       26|    9|   0%| 17|   0%| 15|  9|  1|  4|  0|  0.03  |   2 |   8 |  7 49|  5 40|  86.21
       26|    8|   0%| 18|   0%| 10|  4|  0|  6|  0|  0.02  |   4 |  13 |  7 49|  5 33|  75.00
       25|   11|   0%| 14|   0%| 12|  2|  0| 10|  0|  0.06  |   2 |  25 |  7 49|  2 13|  46.15
       25|   12|   0%| 13|   0%| 14|  5|  0|  9|  0|  0.05  |   1 |  24 |  7 49|  3 20|  72.73
       26|   13|   0%| 13|   0%| 11|  3|  1|  6|  0|  0.02  |   5 |  10 |  7 49|  5 39|  87.88
```

Figure 13.3 Result of detailed measurement on the Abis-interface.

automatic analysis already contains. Figure 13.2a presents the data from the perspective of the BSC, while Figure 13.2b presents the data for the BTSs of that BSC. Figure 13.3 presents a measurement example for the Abis-interface. Table 13.1 explains Figure 13.3.

Table 13.1
Explanation of Figure 13.3

Parameter	Explanation
act	The total number of CHAN_ACT messages per time slot.
assg	The total number of ASS_CMD messages per time slot.
acpl	The success rate for TCH assignment ((ASS_CMP) / (ASS_CMD)).
ho	The total number of HND_CMD messages per time slot.
hcpl	The success rate for TCH assignment ((HND_COM) / (HND_CMD)).
cf	The total number of all CONN_FAIL messages per time slot (all *causes*).
02	The total number of all CONN_FAIL messages per time slot with cause value 02 = *Handover Access Failure*.
1f	The total number of all CONN_FAIL messages per time slot with cause value $1F_{hex}$ = *Radio Link Failure (RLF) Warning*.
28	The total number of all CONN_FAIL messages per time slot with cause value 28 = *Remote Transcoder Failure*.
ei	The total number of ERROR_IND messages per time slot (all *causes*).
Tall	Indicates the average channel occupation time.
Dlev	The average level, with which the MS has received the BTS [dB_m].
Ulev	The average level, with which the BTS has received the MS [dB_m].
Dqul	The average receiving quality on the side of the MS (smaller numbers indicate a better quality).
Uqul	The average receiving quality on the side of the BTS (smaller numbers indicate a better quality).
SACCH%	Fraction of the MEAS_RES messages without MEAS_REP (that is, without measurement results from the MS). The smaller the number, the better.

The company Wandel & Goltermann takes a similar approach with its protocol analyzer MA-10. All the measurement results can be postprocessed, using commercial PC spreadsheet software. Creation of all kinds of statistics and graphical presentations is greatly simplified.

Figure 13.4 presents, as an example, the graphical analysis of an MA-10 trace file captured on the Abis-interface. It shows the graph of the RXLEV value over time for a single call, whereby the measurements for neighboring cells also are shown. The values are taken from MEAS_RES messages. Compare this

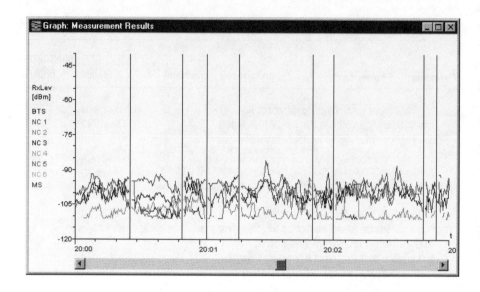

Figure 13.4 Graphical evaluation by means of the protocol analyzer MA-10.

form of analysis with the time-consuming analysis of single messages, for example, when interference needs to be analyzed in the uplink or the downlink of individual TRXs or TSs.

13.2.2 Manual Analysis of Protocol Traces

During the manual analysis of captured signaling data, experienced technicians investigate problems by searching specifically for error messages or inconsistencies, which allows them to draw conclusions on the possible reasons for an unknown problem. The result of such an investigation depends largely on the experience and the ability of the technicians to operate the equipment well.

In contrast to the automatic analysis of measurements on signaling, which also can be used for statistical purposes, the manual approach is more time consuming and hence used only in concrete error situations. Such cases typically are related to errors that cannot be solved by automatic analysis. Protocol errors and timer problems are two examples of such problems.

First of all, it is important during manual analysis to know whether the data are coded in hexadecimal or as normal text. It is much more difficult and more time consuming to analyze hexadecimal data. To make the task easier, previous chapters provided a detailed description of the binary and hexadecimal formats of OSI Layers 2 and 3. That makes those chapters a practical reference. If protocol test equipment is available, the task of decoding is not necessary and interpretation of the data can start immediately. But even in that case, the task of the field engineer still is not easy. By taking advantage of all filter capabilities and search functionality, the essential information has to be extracted. Besides, a thorough understanding of the various scenarios and protocols is mandatory.

13.3 Tips and Tricks

During manual analysis of signaling data, some questions come up repeatedly. This section is intended to provide some help for such questions.

13.3.1 Identification of a Single Connection

13.3.1.1 Connection-Oriented SCCP Mode

The A-interface uses the connection-oriented SCCP mode for all major scenarios. The various messages between BSC and MSC, which are related to an individual connection, can be extracted by using SLR and DLR as a filter. Figure 13.5 illustrates that relationship. The SPC of SCCP network elements (the MSC and BSC, in the A-interface) are not suited, since this value is the same for all messages.

13.3.1.2 Connectionless SCCP Mode

The connectionless SCCP mode is used on the A-interface, for example, to administer the A-interface resources and for paging. There exists hardly any need for filtering connectionless messages. If, however, that should become necessary, the best approach is to use the BSSAP message type as the filter criterion.

Furthermore, the complete MAP communication of the NSS is performed via the connectionless SCCP mode. Because of the fact that MAP scenarios positively are dialogs between subsystems, there is a need to identify all messages of a single dialog. As shown in Figure 13.6, the originating transaction identifier and the destination transaction identifier of the TCAP protocol are well suited for that task.

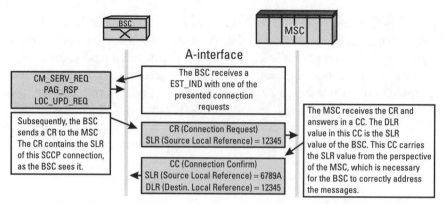

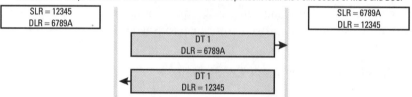

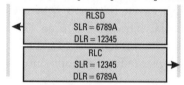

Figure 13.5 Identification of a connection by source local reference and destination local reference.

13.3.1.3 On the Abis-Interface

Before taking measurements on the link between BTS and BSC, the channel configuration has to be determined, since every TRX communicates with the BSC over another TS on the Abis-interface. Otherwise, the risk of capturing the wrong data exists.

A single connection on the Abis-interface can be determined by analyzing the parameter channel number, where a distinction has to be made between CCHs and TCHs (see Chapter 6).

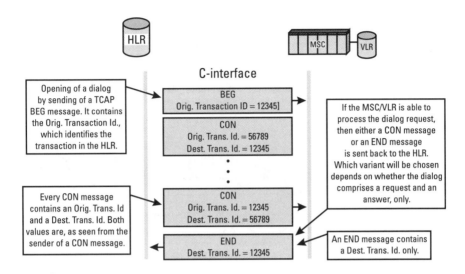

Figure 13.6 Identification of a single MAP connection.

13.4 Where in the Trace File to Find What Parameter?

In which messages and what parameters can the key information be found, and how can the dependencies be detected rapidly? Table 13.2 provides some hints, while Table 13.3 lists statistical information.

13.5 Detailed Analysis of Errors on Abis-Interface and A-Interface

Protocol analyzers are preferably used on the A-interface where it is possible to observe the quality of an entire BSS. The disadvantage of this approach is that it frequently requires additional measurements on the Abis-interface, because the root cause of an error cannot be localized just by looking at the messages on the A-interface.

It is frequently possible, however, because of an unambiguous relationship between error messages on the A-interface and the Abis-interface, to narrow a problem down to a few possible causes within the BSS by analyzing an A-interface trace only. The same applies to an even greater extent to the Abis-interface and the Air-interface. Because the BTS can be regarded as a transparent node for messages between BSC and MS, every error message on the

Table 13.2
General Information

Wanted Information	Interface, Protocol	Parameter, Message Type
Identity of the BTS	A, Abis	Cell Identity (CI) in CM_SERV_REQ, PAG_RSP, LOC_UPD_REQ.
Subscriber identity	A, Abis	IMSI/TMSI in CM_SERV_REQ, PAG_RSP, LOC_UPD_REQ.
Actual LAC during LU	A only	LAI parameter in the *Complete Layer 3* Information message (CL3I).
Former LAC during LU	A, Abis	LAI parameter in the LOC_UPD_REQ message.
Sender/receiver of a SCCP message	SCCP	*Called/Calling Party Address parameter* (Cd/CgPA) in the header of a SCCP message. The Cd/CgPA consists of a combination of a point code, a subsystem number (HLR, VLR, MSC, etc.), and a Global title.
Are there any SCCP problems?	SCCP	Look for CREF messages and UDTS messages. If either message can be found, problems are certain (overload?).
		Check also if all CR's (Connection Request) are answered with CC (Connection Confirm).
Are there any SS7 problems?	SS7	Look for LSSU's and COO's (change over orders). If LSSU's (SIPO or SIB) are detected, then severe SS7 problems on one of the two ends of the SS7 link exist.
Are there any SS7 problems because of high bit error rates?	SS7, OMC	Check if there have been frequent link failures recently. If so, find out if the cause SUERM threshold exceeded is indicated. Look for LSSU's (SIO and SIOS) in the trace file.
Are there any problems in the VLR/HLR?	A, Abis	Look for LOC_UPD_REJ, CM_SERV_REJ, and AUTH_REJ messages. Suspicious causes are: IMSI unknown in HLR, IMSI unknown in VLR, and LAC not allowed. If this occurs frequently, then data errors in the NSS database are likely.
Is there any MS activity in a BSC or a BTS?	A, Abis	Look for CM_SERV_REQ, PAG_RSP, and LOC_UPD_REQ. Detection of CHAN_RQD or IMM_ASS_CMD is not sufficient.

Table 13.2 (continued)

Wanted Information	Interface, Protocol	Parameter, Message Type
Are there Layer 1 problems on the Air-interface?	Abis	Look for CONN_FAIL messages (cause: '1' = Radio Interface Failure). If this occurs frequently, then further investigation is necessary (e.g., identify affected TRX).
Are there Layer 2 problems on the Air-interface?	Abis	Look for ERR_IND messages (frequent cause: '1' = timer T200 expired (N200 + 1) times; 'C_{hex}' = frame not implemented).
Is there interference in the uplink or downlink?	Abis	The RX_QUAL values are poor despite good or acceptable RX_LEV values in the uplink/downlink, frequent intra-BTS handover. Check assignment rate.
Are there problems when sending TRAU frames between transcoder, BTS, and MS	A, Abis	Abis-interface: Look for CONN_FAIL messages (cause: '28hex' = Remote Transcoder Alarm). A-interface: Look for CLR_REQ messages (cause: '20' = Equipment Failure).
Are there problems during incoming handover?	A, Abis	Abis-interface: Look for CONN_FAIL messages (cause: '2' = Handover Access Failure). A-interface: Look for CLR_REQ messages (cause: '0' = Radio Interface Failure).
Are there problems during outgoing handover?	A, Abis	A-interface: Look for HND_FAIL messages. Abis-interface: Look for HND_FAI messages.
Errors in the neighborhood relations? Poor coverage?	A, Abis	Check if there is hardly any outgoing handover. Check if the number of CLR_REQ cause: '1' = Radio Interface Failure (A) and CONN_FAIL (cause: '1' = Radio Link Failure (Abis)) is higher than normal (location dependent).
Are there problems related to interworking between MSC and BSC?	A	High ASS_FAI rate. Causes: Requested Terrestrial Resource unavailable, Terrestrial Circuit already allocated, Protocol Error BSC/MSC. Check trunk assignment and other settings in MSC and BSC. Were the BLO messages, possibly after a reset procedure, not repeated?
Are there any PLMN interworking problems?	MAP	Many ABT messages from the affected PLMN (cause: Application Context Name not supported).

Table 13.2 (continued)

Wanted Information	Interface, Protocol	Parameter, Message Type
Are there any BSC problems?	A	Though the related BTS's do not suffer overload, there are many ASS_FAI messages cause: '33' = Radio Resource unavailable.
Is a BTS blocked?	Abis	Check the RACH control parameters in the SYS_INFOS BCCH_INFOS 1–4. Is the Cell Barr Access bit = 1 or the Access Control Class not equal 0?
MSISDN /IMSI combination of a subscriber	MAP	The BEG/provideRoamingNumber message possibly contains both parameters.
		Another possibility is the BEG/sendRoutingInformation message contains the MSISDN and the
		END/sendRoutingInformation message contains the IMSI.
IMSI /TMSI combination of a subscriber	A only	PAGING message (works on the A-interface, only)
Signaling Point Codes	SS7	Routing Label in every message signal unit (MSU)
Distance between MS and BTS	Abis	Access delay in CHAN_RQD, timing advance (TA) in CHAN_ACT and all MES_RES. For a conversion from TA to distance refer to the Glossary.
Target cell during handover	A	Cell Identity in HND_RQD messages
MS power class (Handy, …)	A, Abis	Mobile Station Classmark X (RF Power Capability) parameter in CM_SERV_REQ, PAG_RSP, LOC_UPD_REQ
Called directory number in case of a MOC	A, Abis	Parameter Called Party BCD Number in SETUP message
Is DTX active?	Abis	DTX (uplink): downlink measurements (MEAS_REP)
		DTX (downlink): uplink measurements (MEAS_RES)

Air-interface translates uniquely into an error message on the Abis-interface; hence, for the majority of the applications, a protocol analysis on the Air-interface is not required.

Table 13.3
Statistical Information

Wanted Information	Interface, Protocol	Parameter, Message Type
Total of all moc attempts (bts/bsc)	Abis, A	$\sum(\text{CM\_SERV\_REQ})$
Total of all mtc attempts (bts/bsc)	A, Abis	$\sum(\text{PAG\_RSP})$
Total of the successful incoming handover	A only	$\sum(\text{HND\_CMP})$
Total of the successful outgoing handover	A only	$\sum(\text{CLR\_CMD [cause: '0B'} = \text{Handover successful]})$
Success rate for MOC's (BSS/BTS)	A, Abis	$\dfrac{\sum(\text{ALERT [from MSC} \to \text{MS]}) + \sum(\text{PROGRESS})}{\sum(\text{CM\_SERV\_REQ [Establishment cause} = \text{MOC]})}$
Error rate for MOC's (BSS/BTS)	A, Abis	$1 - \dfrac{\sum(\text{ALERT [from MSC} \to \text{MS]}) + \sum(\text{PROGRESS})}{\sum \text{CM\_SERV\_REQ [Establishment cause} = \text{MOC]}}$
Success rate for MTC's (BSS/BTS)	A, Abis	$\dfrac{\sum(\text{ALERT [from MS} \to \text{MSC]})}{\sum(\text{PAG\_RSP})}$
Error rate for MTC's (BSS/BTS)	A, Abis	$1 - \dfrac{\sum(\text{ALERT [from MS} \to \text{MSC]})}{\sum(\text{PAG\_RSP})}$
Success rate for incoming handover	A only	$\dfrac{\sum(\text{HND\_CMP})}{\sum(\text{HND\_REQ})}$
Error rate for incoming handover	A only	$1 - \dfrac{\sum(\text{HND\_CMP})}{\sum(\text{HND\_REQ})}$
Success rate for outgoing handover	A only	$\dfrac{\sum(\text{CLR\_CMD [cause: '0B'} = \text{handover successful]})}{\sum(\text{HND\_CMD})}$
Error rate for outgoing handover	A only	$1 - \dfrac{\sum(\text{CLR\_CMD [cause: '0B'} = \text{handover successful]})}{\sum(\text{HND\_CMD})}$

13.5.1 Most Important Error Messages

13.5.1.1 Clear Request (CLR_REQ) on the A-Interface

A CLR_REQ indicates a connection error detected in the BSS and mandates an abnormal termination. Except for error indications during the assignment procedure, the CLR_REQ message is used for various error scenarios, including incoming and outgoing handover, as well as radio link failures in Layers 1 and 2.

In particular, CLR_REQ messages with cause '0' = radio interface message failure and cause '1' = radio interface failure frequently can be found in trace files.

The difference between the causes is that a *radio interface message failure* is reported when, during a Layer 3 scenario, signaling messages from the MS are not received on time. For example, during a mobile originating call setup, the SETUP message from the mobile station is not received. The connection is torn down by the BSS with CLR_REQ and cause '0' = radio interface message failure. On the other hand, CLR_REQ with cause '1' = radio interface failure is indicated, when problems in Layers 1 and 2 were detected on the Air-interface that require abnormal termination of a connection. Note that a *radio interface message failure* occurs most likely during connection setup, when Layer 3 signaling data need to be exchanged, while a *radio interface failure* can happen at any time during an active connection. In both cases, radio interface problems are the actual problem source.

As presented in Figure 13.7, the MSC reacts on receiving a CLR_REQ message by sending a CLR_CMD message to release the still seized radio resources.

13.5.1.2 Assignment Failure on the A-, Abis-, and Air-Interfaces

In contrast to the CLR_REQ, which can be sent any time during an active connection, the ASS_FAIL indication can frequently be sent only as a reaction to

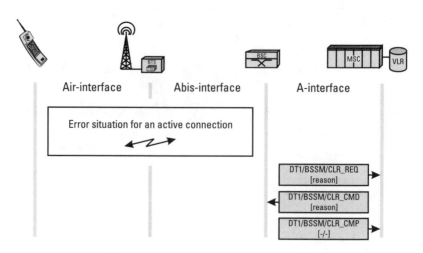

Figure 13.7 Use of the CLR_REQ message.

a faulty traffic channel assignment during call establishment. In that case, one has to distinguish between ASS_FAI on the Air-interface and Abis-interface, as defined in GSM 04.08, and the ASS_FAIL on the A-interface, as defined in GSM 08.08. The ASS_FAI is used on the Air-interface and the Abis-interface as a negative response for an ASS_CMD message, while ASS_FAIL is used as a reaction for an ASS_REQ message on the A-interface. Note that every ASS_FAI/ASS_FAIL message indicates an aborted connection setup. For that reason, network operators react sensitively on an increasing ASS_FAI/ASS_FAIL rate.

Another reason to use the ASS_FAI/ASS_FAIL rate as a rough indicator for the network quality lies in the fact that for the BSS and the radio interface, the assignment procedure is the most critical phase during call establishment. Any radio or BSS related problems most likely will become visible during the assignment procedure.

If the assignment procedure is unsuccessful, one of the following scenarios applies (see Figure 13.8):

- The BSC may reject an ASS_REQ, for instance, if the MSC tries to assign a channel on the A-interface that the BSC regards as not available or that is already in use. In such cases, ASS_FAIL with the respective cause is returned to the MSC, without the BSC sending a ASS_CMD message to the MS. Such an error cannot be investigated on the Abis-interface.

- When the BSC is able to process the ASS_REQ, it assigns a free traffic channel on the Air-interface and forwards the corresponding information in an ASS_CMD message (via the BTS) to the MS. The BSC then

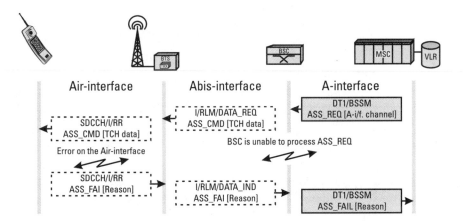

Figure 13.8 ASS_FAIL/ASS_FAI on A-, Abis-, and Air-interface.

waits for a defined time period, as specified by BSS timer T10, for a response from the MS.

- If the MS is unable to seize that traffic channel, it tries to send an ASS_FAI message with the appropriate RR cause over the SDCCH. Only in those cases will the ASS_FAIL message on the A-interface contain an RR cause.
- If the MS, because of an interrupted radio connection, is unable to send either an ASS_COM or an ASS_FAI to the BSC, the BSS timer T10 expires and an ASS_FAIL message with cause '0' = radio interface message failure is sent over the A-interface to the MSC.

13.5.1.3 Connection Failure (CONN_FAIL)

CONN_FAIL is an error message on the Abis-interface that indicates a Layer 1 error on the Air-interface. The most frequent reasons for a CONN_FAIL message are the following:

- The radio connection is lost on the uplink or downlink. In GSM, this kind of error is referred to as radio link failure (see Figure 13.9). After the BTS detects that the connection is lost, it sends a CONN_FAIL with cause '451' = radio link failure to the BSC. Radio link failures occur frequently on the SDCCH (particularly between AUTH_REQ and AUTH_RSP, since there is a fairly long time between the two), because a poor radio connection could just be established but not stay stable. For that reason, during a statistical analysis, it is advised that there be a distinction between errors on the SDCCH and errors on the TCH.
- Errors during incoming handover (cause '2' = handover access failure) (see Figure 13.10). The access on the new channel does not work and the BTS informs the BSC by sending a CONN_FAIL message. The occupied resources are released.
- Error during the TRAU frame synchronization in the uplink or the downlink after assignment of a traffic channel. In most cases, the problems are in the uplink direction, for example, because the wrong channel type was assigned (halfrate instead of fullrate, DTX on/off). When this error is detected in the TRX, the BTS sends a CONN_FAIL with cause '28_{hex}' = remote transcoder failure to the BSC (see Figure 13.11). Note that sporadic CONN_FAIL (remote transcoder failure) may occur, even without systematic errors. The rate should, however, not exceed 1.0% of all call attempts.

Quality of Service 295

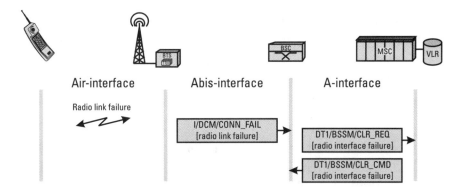

Figure 13.9 CONN_FAIL (cause '1' = radio link failure).

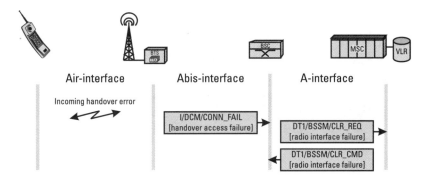

Figure 13.10 CONN_FAIL (cause '2' = handover access failure).

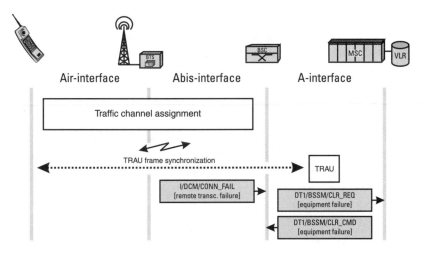

Figure 13.11 CONN_FAIL (cause '28_{hex}' = remote transcoder failure).

13.5.1.4 Error Indication (ERR_IND)

The ERR_IND message indicates a protocol error in Layer 2 on the Abis-interface, which occurred on the Air-interface. The reason, however, typically is not a protocol error but problems with the transmission as a consequence of a poor radio connection. Consequences of a poor radio connection may be the following:

- An increased bit error rate, which in turn can affect the $LAPD_m$ protocol. Examples include wrong frame type and P/F bit or C/R bit set wrong. In many of those cases, an ERR_IND with cause '$0C_{hex}$' = frame not implemented is sent to the BSC. When the BSC receives an ERR_IND message with this cause, it takes no special action, in particular, it does not tear down the connection (Figure 13.12).

- Repeated sending of $LAPD_m$ frames, for which the peer expects an acknowledgment, which it does not receive (Figure 13.13). If a Layer 2 message was repeated without acknowledgment as many times as indicated by the parameter N200, then an ERR_IND with cause '1' = timer T200 expired (N200 + 1) times is sent to the BSC. The BSC will then try to release the Layer 2 and Layer 1 resources, which in turn mainly results in a radio link failure in Layer 1, because the MS is not there anymore (see *Radio Link Failure* and the values for N200 in the Glossary).

13.5.2 Error Analysis in the BSS

This section presents hints on how to investigate the most frequent errors in a GSM network. The general approach is the exclusion method, that is, eliminating possible causes one by one, which is applied by many "cookbooks" that deal with this subject. Note that the explanation provides a good help to solving

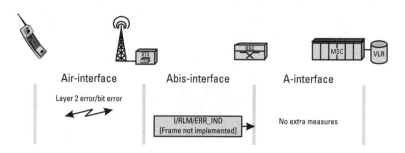

Figure 13.12 ERR_IND / Cause '$0C_{hex}$' = Frame not implemented.

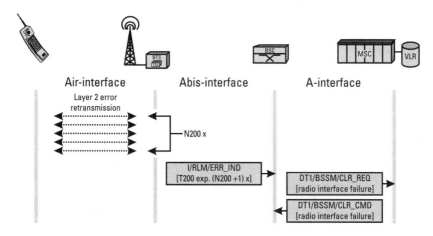

Figure 13.13 ERR_IND (cause '01' = timer T200 expired (N200 + 1) times).

many problems, but it is by no means the universal approach that always leads to success. Most times, field engineers need only an initial hint to be brought on the right track; from there, they are able to narrow the problem down on their own.

Also note that some of the proposed actions may be performed only during low traffic hours and only with the agreement of the network operator.

13.5.2.1 CLeaR REQuest (cause value: '0' = Normal Event—Radio Interface Message Failure)

An increased number of CLR_REQs point to problems during outgoing intra-MSC handover and during inter-MSC handover. In case of such a scenario, after sending a HND_CMD message, the old BSC expects a CLR_CMD message from the MSC within the time period defined by the BSS timer T8, to release the still occupied radio resources (Figure 13.14). If the old BSC receives no CLR_CMD or no HND_FAIL from the MS/BTS, it sends a CLR_REQ with the cause, radio interface message failure, to the MSC to trigger the release the occupied resources.

13.5.2.2 CLeaR REQuest (Cause Value: '1' = Normal Event—Radio Interface Failure)

A higher number of this CLR_REQ points toward problems during connection setup, during a connection, or during incoming handover. The cause for this error typically is related to the BTS or the radio link. Frequently, a CLR_REQ with cause '1' is the reaction for a radio link failure on the Air-interface (CONN_FAIL with cause '1' = radio link failure on the Abis-interface) or

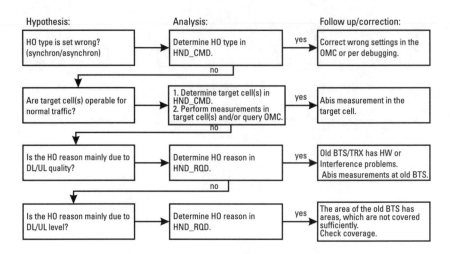

Figure 13.14 Investigation of CLR_REQ/cause value: '0' = Normal event-Radio Interface Message Failure.

a reaction for errors during channel seizure during an incoming handover (CONN_FAIL with cause '2' = handover access failure on the Abis-interface)(Figure 13.15).

13.5.2.3 CLeaR REQuest (Cause Value: '20'hex = Resources Unavailable—Equipment Failure)

CLR_REQs with this cause, if they occur in a higher number, to problems with the connection between transcoder and the BTS (Figure 13.16). A CLR_REQ with the cause of equipment failure on the A-interface frequently is the result of a CONN_FAIL with cause '28'hex = remote transcoder alarm on the Abis-interface.

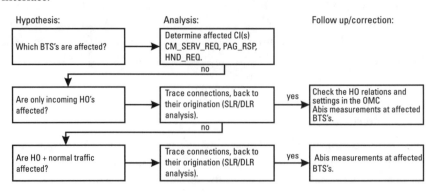

Figure 13.15 Investigation of CLR_REQ/cause value: '1' = Normal event-Radio Interface Failure.

Quality of Service

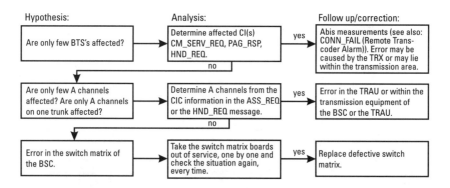

Figure 13.16 Investigation of CLR_REQ/cause value: '20'$_{hex}$ = Resources unavailable-Equipment Failure.

13.5.2.4 ASSignment FAILure (Cause Value: '50'hex Invalid Message—Terrestrial Circuit Already Allocated)

This case of ASS_FAIL should be taken seriously, even in case of a few occurrences. This cause is sent when inconsistencies exist between BSC and MSC about the state of A channels. Different from the ASS_FAIL (Resources Unavailable—Requested Terrestrial Resource Unavailable), this case of ASS_FAI (Invalid Message—Terrestrial Circuit Already Allocated) is sent if the BSC determines that the channel the MSC has requested is—although generally available for traffic—already occupied (Figure 13.17).

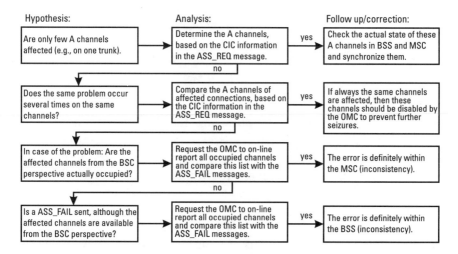

Figure 13.17 Investigation of ASS_FAIL/cause value: '50'$_{hex}$ = Invalid Message-Terrestrial Circuit already allocated.

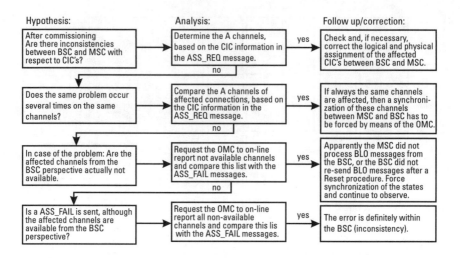

Figure 13.18 Investigation of ASS_FAIL/cause value: '22'$_{hex}$ = Resources unavailable-Requested terrestrial resource unavailable.

13.5.2.5 ASSignment FAILure (Cause Value: '22'hex Resources Unavailable—Requested Terrestrial Resource Unavailable)

This case of ASS_FAIL should be taken seriously, even in case of only a few occurrences. This cause is sent when inconsistencies occur between BSC and MSC about the state of A channels (Figure 13.18). Different from the ASS_FAIL (Invalid Message—Terrestrial Circuit Already Allocated), this ASS_FAIL is sent if the BSC determines that the channel the MSC has selected is not already occupied by another connection but generally is not available for traffic.

13.5.2.6 Infrequent SRES Mismatch Error Messages at the OMC

Description of the Error

One of our customers filed a complaint that the OMC receives an increased number of SRES mismatch error messages. Such error messages are not unusual. They are indicated, for instance, if someone tries to make a phone call with an invalid SIM card. The difference in this case was that all SRES mismatches originated in only one BSC area.55

It could not be completely ruled out that someone used a faulty SIM card in this BSC area only. However, a more detailed analysis showed that even perfect SIM cards were affected by this phenomenon. It could easily be told that the SIM cards were perfect, because they worked flawlessly immediately after the problem occurred. Therefore, a technical defect was suspected, with a high probability of being located in the BSS.

Error Analysis

The analysis was performed on site with a protocol analyzer. For that purpose, all the SS7 connections between MSC and BSC were monitored for several hours, and the signaling data were captured and later analyzed. The tracing period was long enough that several errors could be recorded. The investigation revealed the situation on the A-interface and is illustrated in Figure 13.19.

Affected MSs almost simultaneously ($\Delta t = 20$ ms) sent two identical CM_SERV_REQ messages on the A-interface to the MSC, with a normal establishment cause. The values for SLR and signaling link selection (SLS) were, however, different for the two connection requests. The MSC/VLR responded to the first request with an AUTH_REQ and to the second one with a CM_SERV_REJ with the cause congestion. In line with the protocol, the MS then aborted the connection. Further investigations, carried out simultaneously on the A-interface and various Abis-interfaces revealed that the CM_SERV_REQ message, which was duplicated on the A-interface, was actually only a single message on the Abis-interface. Therefore, the error had to be located in the BSC. It appeared as if the BSC forwarded some messages with

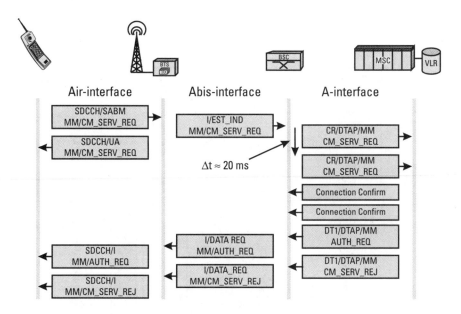

Figure 13.19 SRES mismatches caused by duplicate CM_SERV_REQs.

echo. A detailed analysis of the trace file revealed also that this problem was not restricted to CM_SERV_REQ messages, but other messages were also duplicated.

Error Correction

BSC internal protocol measurements performed by utilizing low-level measurement tools helped to detect a faulty board in the switch matrix, which was exchanged.

13.5.2.7 All TRXs Perform Restarts When Brought Into Service

Description of the Error

During the commissioning of a BSC massive problems occurred. All TRXs in all connected BTSs did not start their service but were continually restarting. A hardware problem was extremely unlikely and could be ruled out. For that reason, an affected Abis-interface was more closely monitored by means of a protocol analyzer.

Error Analysis

The investigation with the protocol analyzer quickly revealed the cause of the error. As Figure 13.20 shows, the RF_RES_IND were not sent in intervals of TX = 120s, as mandated, but with the wrong interval of TX = 0s.

Error Correction

The settings of TX were changed in the BTS-related software in the BSC. All BTSs were loaded again, and the problem was fixed.

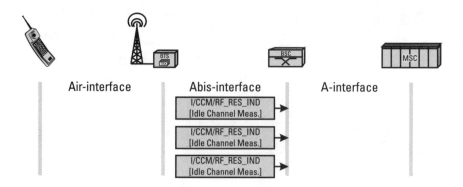

Figure 13.20 TRX restarts caused by permanently repeated RF_RES_IND messages.

Glossary

Anyone who works on GSM issues will encounter many terms and parameters that have specific meanings in the telecommunications environment. This glossary provides an alphabetically ordered description of a significant number of these terms. Many of the descriptions are supplemented with references to GSM and ITU Recommendations, shown in brackets [...].

26-Multiframe *See 51 multiframe.*

51-Multiframe Time slots for transport of information in a GSM system are organized in frames. One TDMA frame consists of 8 time slots, each 0.577 ms long. TDMA frames are organized in multiframes. Two such multiframes are defined, one with 26 TDMA frames (26-multiframe) and one with 51 TDMA frames (51 multiframe). Multiframes are organized in superframes, and superframes are organized in hyperframes. For more details, see Chapter 7.

A-interface [GSM 04.08, 08.06, 08.08] The interface between BSC and MSC. For more details, see Chapters 8, 9, and 10.

A-law [G.711] Spoken language generally is not linear in its dynamics, and the human ear is rather sensitive to soft sounds, but difference in amplitude for loud sounds cannot be distinguished so easily. When digitizing speech, one can take advantage of this situation and code a sufficient-quality sound with relatively few bits. In particular, the relative error that is made when quantizing needs to be minimized. The relative error is $\Delta x/x$ or dx/x. To minimize that value for all cases, it has to be constant. Since the integral of 1 over x equals the natural logarithmic function (as in the equation $\int \frac{dx}{x} = \ln(x) + C$, a logarithmic function best suits that objective. For this purpose, the A-law and the μ-law were invented. Both are approximations of the natural logarithmic function, and both were standardized by ITU for transmission of digital speech on PCM transmission lines, as shown in Figure G.1(a).

Both methods are used on a per-country basis. The μ-law is used only in the United States and Japan. All other countries use the A-law. The international standard G.711 deals with the case of an international connection that involves two countries where different methods are used. The standard requires that, independent of the origination, a possibly necessary transformation be carried out in the country that uses the μ-law.

The first step is the same for both methods, that is, to sample the analog signal with a sampling rate of 8 kHz. The sample then is quantized according to the respective law and coded in 8-bit code words. That results in the transmission rate of 64 Kbps, used on PCM channels. Both methods differ only in a slight variation of the no-linear quantization of the sample.

Figure G.1(b) is a graphic representation of the A-law, and Figure G.1(c) provides the representation of the μ-law. The first bit indicates whether the value is positive or negative, the following 3 bits define the segments, while the bits marked with "x" represent values within that segment.

A3, A5/X, A8 [GSM 03.20] Names of three algorithms used in GSM for authentication and ciphering (Figure G.2). All the algorithms used in GSM are highly confidential and therefore not published in any standard.

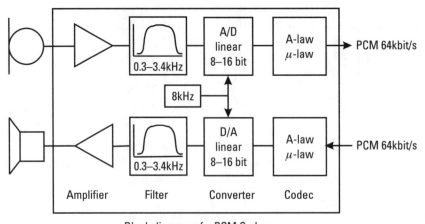

Block diagram of a PCM Codec

Figure G.1(a) A-law and μ-law for digitalization of speech.

Glossary

Segment	Code word		Segment
7	1111xxxx	1V	
6	1110xxxx	1/2V	
5	1101xxxx	1/4V	
4	1100xxxx	1/8V	
3	1011xxxx	1/16V	
2	1010xxxx	1/32V	
		1/64V	
1	10011111 10000000	0V	1
		00000000 00011111	
	−1/64V	0010xxxx	2
	−1/32V	0011xxxx	3
	−1/16V	0100xxxx	4
	−1/8V	0101xxxx	5
	−1/4V	0110xxxx	6
	−1/2V	0111xxxx	7
−1V			

Figure G.1(b) Graph for the A-law.

Segment	Code word		Segment
8	1000xxxx	1V	
7	1001xxxx	1/2V	
6	1010xxxx	1/4V	
5	1011xxxx	1/8V	
4	1100xxxx	1/17V	
3	1101xxxx	1/36V	
2	1110xxxx	1/86V	
		1/264V	
1	1111xxxx	0V	1
	−1/264V	0111xxxx	2
	−1/86V	0110xxxx	3
	−1/36V	0101xxxx	4
	−1/17V	0100xxxx	5
	−1/8V	0011xxxx	6
	−1/4V	0010xxxx	7
	−1/2V	0001xxxx	
−1V		0000xxxx	8

Figure G.1(c) Graph for the μ-law.

The "X" in A5/X indicates that there are several A5 algorithms. The network and the mobile station (MS) have to agree on one of these algorithms before ciphering can be used. The MS does not necessarily "know" every algorithm. Originally, GSM had only one algorithm, A5, but due to export restrictions of security codes, more less-secure algorithms were defined. The algorithm A5 is built into the MS, not into the SIM. GSM has defined A5/1 through A5/7, and the MS uses an information element, the mobile station classmark, to inform the network during connection setup which algorithms it actually supports.

Abis-interface [GSM 04.08, 08.58]   The interface between BTS and BSC. For more details, refer to Chapter 6.

Access class   GSM recognizes 16 different access classes. This parameter is stored on the SIM module and allows the network operator to specifically bar certain types of subscribers. A typical application is to set up an access class exclusively for the operator personnel for test purposes during installation and testing. In that case, the system can be on the air but ordinary users do not recieve access. Another application is to define access classes for emergency personnel only. This can prevent overload during an emergency and allows rescue workers to be reachable via mobile phone.

The BTS broadcasts the admitted access classes within the RACH control parameters, which are part of the information that the BTS permanently broadcasts in its broadcast control channel (BCCH). The MS reads the information and compares it with the access classes on the SIM. The MS attempts to access the system only if it finds a matching access class. That prevents signaling overload because an unauthorized MS does not even try to access the system.

Table G.1
Application of the GSM Algorithms A3, A5/X, and A8

Algorithm	Dependency	Remark
A3	SRES = f(A3, K_i, RAND)	The MS calculates the SRES by using the RAND as a parameter for the A3 algorithm.
A5/X	CS = f(A5/X, K_c, FN)	MS and BTS both need the ciphering sequence for the ciphering process.
A8	K_c = f(A8, K_i, RAND)	K_c is calculated from A8, K_i, and RAND. It is then used as an input parameter for ciphering.

The access classes in GSM use values from 0 to 15. The numbers do not indicate any priority as such, that is, a higher number does not imply a higher priority or vice versa. Table G.2 shows the use of the access classes. "Ordinary" subscribers receive values from 0 through 9 on a random basis. Only the access classes 11 through 15 were predefined. Note that one SIM module is capable of storing several access classes, which allows one subscriber to belong to several subscriber groups.

Access delay Synonym for timing advance (TA).

ACCH [GSM 05.01, 05.02] Associated control channel. Two types are defined: slow associated control channel (SACCH) and fast associated control channel (FACCH). An ACCH is assigned for traffic channels (TCHs) as well as for SDCCHs.

Adjacent cells See *Neighbor cell*.

AE [GSM 09.02, X.200–X.209] The term application entity (AE) is used by the OSI Reference Model in which it refers to a physical entity in Layer 7, the application layer. The different protocols for the GSM network elements HLR, VLR, and EIR are examples of AEs. Refer to Chapter 11 for more details about AEs.

AGCH [GSM 05.01, 05.02] Access grant channel. A common control channel (CCCH) that is used only in the downlink direction of even-numbered

Table G.2
Access Classes in GSM

Access Class (Decimal)	Subscriber Group
15	Network operator personnel
14	Emergency service
13	Public services (utilities)
12	Security service
11	To be assigned by the operator
10	Not used
0–9	"Ordinary" subscribers

time slots (typically solely in time slot 0) of the BCCH-TRX. It is used to assign a SDCCH to the MS, and it transports the IMM_ASS (IMMediate ASSign) message. Depending on the chosen channel configuration, the AGCH shares the available downlink CCCHs with the paging channel (PCH) and the SDCCH. The transmission rate per AGCH block is 782 bps. (See Chapter 7.)

Air-interface [GSM 04.XX, GSM 05.XX]   The interface between MS and BTS. In an analogy to the fixed network, this interface is also referred to as the U_m interface. Chapter 7 provides more details.

AIS [G.703]   Alarm indication signal. An alarm known from the transmission systems. A terminal shows the alarm when the Layer 1 connection to the next entity is working properly, but the peer entity still is not reachable because somewhere down the line the connection is broken. For example, consider a BTS that has a connection to a BSC. The connection is composed of two links connected in serial, as shown in Figure G.2. If one of the two connections fails, for example, the one next to the BSC, the BTS shows an AIS.

AoC/AoCI/AoCC   Three terms relative to the GSM charging type of supplementary services (SS). Advice of charge (AoC) (the generic term for the two specific SSs), advice of charge indication (AoCI), and advice of charge charging (AoCC). The difference lies in the level of accuracy. For more details, see Supplementary services.

APDU   See *PDU.*

Application context name [GSM 09.02, X.208, X.209]   An identifier used by the transaction capabilities application part (TCAP) that identifies which protocol an application has to use. Optional information element of the dialog part in a TCAP message.

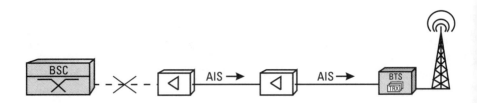

Figure G.2 Use of the alarm indication signal (AIS).

The application context in the GSM-MAP identifies the application to be used for execution of a MAP dialog in the HLR, VLR, MSC, or EIR. More details are provided in Chapter 11.

Application entity   See *AE*.

ARFCN [GSM 05.01]   Absolute radio frequency channel number. An identifier or number of a channel used on the Air-interface. From the ARFCN, it is possible to calculate the frequency of the uplink and the downlink that the channel uses. How to perform this calculation is shown under downlink.

ASE [GSM 09.02, X.200–X.209]   Application service element. Single-user protocol of OSI Layer 6. For example, the whole GSM MAP (Layer 7) is an ASE of the transaction capabilities application part (TCAP), while individual parts of MAP (e.g., HLR, VLR) are referred to as application entities (AEs).

ASN.1 [Q.771, Q.772, Q.773, X.208, X.209]   Abstract Syntax Notation number 1 (ASN.1) is the first—and so far only—standardized means to describe operations of interfaces and their parameters. ITU has standardized this notation in its Recommendations X.208 and X.209, based on the OSI Reference Model (X.200 through X.207).

The interfaces of the mobile application part (MAP) of GSM are specified with the use of ASN.1.

An important part of ASN.1 is the definition of how to assign parameter identifiers, depending on their category and the type of application. A parameter identifier is called TAG.

Table G.3 shows the encoding of the various parameter types defined in X.208. Encoding of those bits, indicated by X, are determined by the type of application. For more details, see Chapter 11.

ATT   See *IMSI attach, IMSI detach; BCCH_INFO SYS_INFO 1–4*.

AuC [GSM 03.02, 03.20]   Authentication center. Part of the network switching subsystem (NSS). It is a physical part of the HLR. For more details, see Chapter 4.

Authentication [GSM 03.20]   Getting access to telecommunication services by cloning of a valid user identifier is a common problem in many mobile networks. GSM anticipated that problem and defined an authentication procedure: an operation that prevents unauthorized use of service by challenging a user to provide proof of the claimed identity. After the user requests access to

Table G.3
Parameter Types in ASN.1

Encoding								Parameter Type
7	6	5	4	3	2	1	0	Bit
X	X	0	0	0	0	0	1	Boolean
X	X	0	0	0	0	1	0	Integer
X	X	0	0	0	0	1	1	Bitstring
X	X	X	0	0	1	0	0	Octetstring
X	X	X	0	0	1	0	1	Null
X	X	X	0	0	1	1	0	Object identifier
X	X	X	0	0	1	1	1	Object descriptor
X	X	X	0	1	0	0	0	External
X	X	X	0	1	0	0	1	Real
X	X	X	0	1	0	1	0	Enumerated
X	X	X	1	0	0	0	0	Sequence/sequence of
X	X	X	1	0	0	0	1	Set/set of
X	X	X	1	0	0	1	0	Character string
X	X	X	1	0	0	1	1	Character string
X	X	X	1	0	1	0	0	Character string
X	X	X	1	0	1	0	1	Character string
X	X	X	1	0	1	1	0	Character string
X	X	X	1	1	0	0	1	Character string
X	X	X	1	1	0	1	0	Character string
X	X	X	1	1	0	1	1	Character string
X	X	X	1	0	1	1	1	Time
X	X	X	1	1	0	0	0	Time

the network and provides the user identifier, the network sends a random number (RAND) to the MS. The input is used, together with some secret information on the SIM and a secret algorithm, to provide a response (SRES). More details can be found under ciphering.

B-interface [GSM 09.02] The interface between MSC and VLR. Since the time when GSM Phase 2 was specified, this interface is no longer part of the external interfaces, and SMG provides no detailed specifications. For more details, see Chapter 4.

BAIC, BAOC, BIC Roam, BOIC, BOICexHC   Supplementary services (SS) that bar certain types of calls: Barring of all incoming calls (BAIC), barring of all outgoing calls (BAOC), barring of incoming calls while roaming (BIC-Roam), barring of outgoing international calls (BOIC), barring of outgoing international calls except those to the home country (BOICexHC). For more details, see SS.

Bbis [GSM 04.06]   Frame format used on the Air-interface for the $LAPD_m$ protocol exclusively to transmit the BCCH, PCH, and AGCH. It is different from the regular $LAPD_m$ frame format in that Bbis utilizes neither address or control fields nor length indicators. For more details, refer to Chapter 7.

BCC [GSM 03.03]   Base station color code. A 3-bit-long parameter that is part of the BSIC. Used to distinguish among the eight different training sequence codes (TSCs) that one BTS may use on the CCCHs and to distinguish between neighbor BTSs without the need for the MS to register on any other BTS.

BCCH [GSM 04.08, 05.01, 05.02]   Broadcast common control channel. The "beacon" of every BTS. Per BTS, there is always exactly one BCCH, which is transmitted in time slot 0 of the BCCH frequency. The transmission rate is 782 bps.

BCCH / SYS_INFO 1–4 [GSM 04.08]   Message sent on the BCCH for radio resource management purposes. Several types of this message exist. Types 1 through 4 are explained in more detail.
   A BTS uses the SYS_INFO 1–4 on the BCCH to provide all cell-specific data to every MS that receives the signal. That includes accessibility, available services, neighbor cells, radio frequencies, and so on. The BSC provides the relevant BCCH information to each BTS individually. An example captured from a GSM system in East Asia illustrates the content of the BCCH information (Figures G.3 through G.6). Note that the number of neighbor cells, and hence frequencies in DCS1800 and PCS1900 is large and, therefore, exceeds the capacity of SYS_INFO 2. This large number of frequencies is required to define and broadcast a SYS_INFO 2bis and SYS_INFO 2ter, in addition.
   The following description shows that SYS_INFO 4 only provides repetition of the already sent parameters. Only with active cell broadcast (CB) does

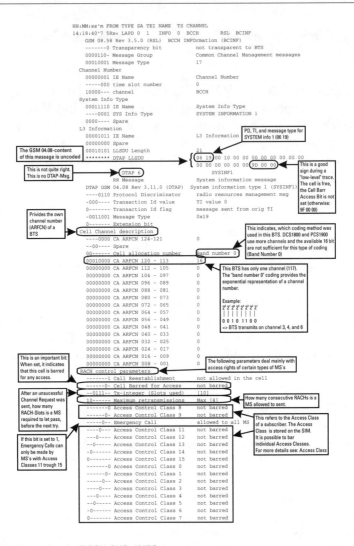

Figure G.3 Example of a BCCH SYS_INFO 1 message.

SYS_INFO 4 provide new information by describing the cell broadcast channel.

SYS_INFO 5 and 6 [GSM 04.08] In contrast to BCCH/SYS_INFO 1–4, which broadcasts all BTS-specific data to the mobile stations in the idle case, the SYS_INFO 5 and 6 perform that task when there is an active connection either on a SDCCH or on a TCH. Note that in DCS 1800 and PCS 1900 the

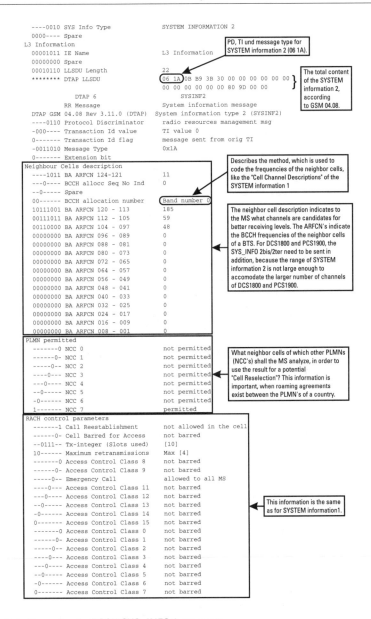

Figure G.4 Example of a BCCH SYS_INFO 2 message.

SYS_INFO 5 was expanded by SYS_INFO 5bis and 5ter, to accommodate the greater number of available frequencies of neighbor cells. More details on the use of SYS INFO 5 and 6 can be found in Chapter 12.

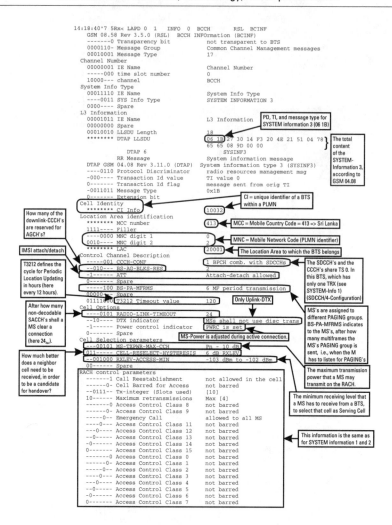

Figure G.5 Example of a BCCH SYS_INFO 3 message.

BCCH SYS_INFO 7 and 8 Specific messages for DCS 1800 and PCS 1900. They share a logical channel with BCCH SYS_INFO 4.

BCD Binary coded decimal. A method to code a decimal digit with four binary bits. If the decimal number has several digits (i.e., because it is greater than 9), each digit is coded individually, for example, 79_{dez} = '0111 1001'$_{bin}$.

Bearer services [GSM 02.01, 02.02] Different transmission capabilities that GSM provides, as listed in Table G.4. Note that bearer services need to be

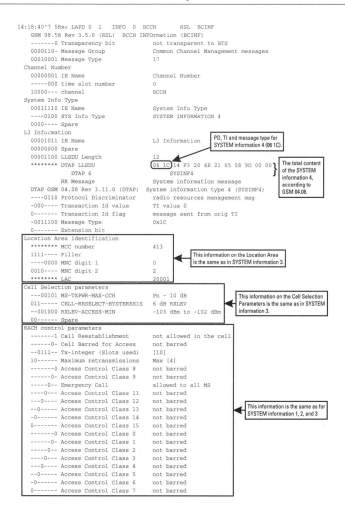

Figure G.6 Example of a BCCH SYS_INFO 4 message.

distinguished from teleservices, which include possibly required terminal equipment.

BER Bit error rate on the Air-interface. Determined by the value of RXQUAL.

BERT Bit error rate test. A measurement of bit error rates.

BFI [GSM 05.05, 06.31, 08.60] Bad frame indicator. A parameter within the TRAU frame. The value of the BFI indicates to the voice decoder if a

Table G.4
Bearer Services in GSM

Bearer Service		Remarks
No.	Name	
21	Asynchron/300 baud	-/-
22	Asynchron/1200 baud	-/-
23	Asynchron/1200 baud/75 baud	Only for mobile originating call. 1200 Baud for downlink and 75 baud for uplink.
24	Asynchron/2400 baud	-/-
25	Asynchron/4800 baud	-/-
26	Asynchron/9600 baud	-/-
31	Synchron/1200 baud	-/-
32	Synchron/2400 baud	-/-
33	Synchron/4800 baud	-/-
34	Synchron/9600 baud	-/-
41	Asynchron over PAD 300 baud	Only mobile originating call.
42	Asynchron over PAD 1200 baud	Only mobile originating call.
43	Asynchron over PAD 1200 baud /75 baud	Only mobile originating call. 1200 baud on downlink and 75 baud on uplink.
44	Asynchron over PAD/2400 baud	Only mobile originating call.
45	Asynchron over PAD/4800 baud	Only mobile originating call.
46	Asynchron over PAD/9600 baud	Only mobile originating call.
51	Synchron/packet access/2400 baud	Only mobile originating call.
52	Synchron/packet access/4800 baud	Only mobile originating call.
53	Synchron/packet access/9600 baud	Only mobile originating call.
61	Speech and data	This allows to switch back and forth between speech and data. During the data phase bearer services 21–26 are used.
81	Speech followed by data	After the switch to data transmission, it is not possible to switch back to speech. Bearer services 21–26 are used during the data phase.

TRAU frame contains valid data (BFI = 0) or not (BFI = 1). Depending on that information, the voice decoder uses or discards a TRAU frame. Note: For FACCH frames, BFI always equals 1, because they contain signaling data.

BIB [Q.700–Q.704] Backward indicator bit. Used to detect transmission errors in SS7 messages. Other related indicators are the BSN, FSN, and FIB. BIB is the most significant bit (MSB) of the first byte of an SS7 message (FISU, MSU, LSSU). For more details, see Chapter 8.

B_m channel [GSM 04.03] Another term for the GSM fullrate channel. It allows transmission of speech at a rate of 13 Kbps.

BS Bearer service.

BS_AG_BLKS_RES [GSM 05.02] Parameter transmitted with the BCCH SYS_INFO 3 message. BS_AG_BLKS_RES is 3 bits long and hence can take on the values 0 through 7. The value of this parameter indicates to all mobile stations in a cell how many of the CCCH blocks of a 51 multiframe on a BCCH-TS 0 are reserved for access grant channels (AGCHs). The number of available paging channels (PCHs) is reduced accordingly. Note that during operation in the combined mode of SDCCH and CCCH the number of CCCH blocks per time slot is four rather than eight, compared to the non-combined mode. The complete picture is illustrated in Figure G.7. (See also CCCH_CONF).

BS_PA_MFRMS [GSM 05.02] Mobile stations are organized into paging groups based on their IMSI. A mobile station that belongs to a certain paging group needs to check for a paging message only once in a number of 51 multiframes. In between, the mobile station may switch over to an energy-saving mode, discontinuous reception (DRX).

The 3-bit-wide parameter BS_PA_MFRMS is part of the BCCH SYS_INFO 3 and tells the mobile station after how many multiframes the content of the paging channel (PCH) has to be analyzed by the MS. In other words, this parameter indicates how often a particular paging group is repeated. Figure G.8 provides an example of how this parameter is used.

BS_CC_CHANS [GSM 05.02] Parameter that indicates how many time slots on the BCCH frequency are reserved for common control channels

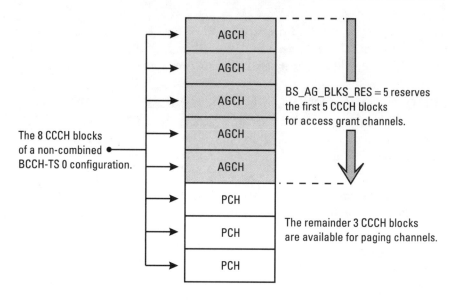

Figure G.7 The meaning of BS_AG_BLKS_RES.

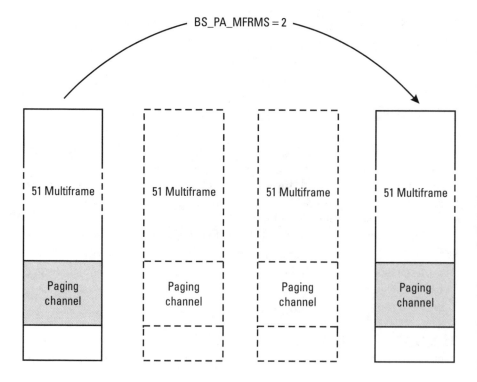

Figure G.8 The task of the BS_PA_MFRMS.

(CCCHs). This parameter is not transmitted but is derived from another parameter, CCCH_CONF.

BS_CCCH_SDCCH_COMB [GSM 05.02] Parameter that indicates whether the dedicated control channels (SDCCHs) and the common control channels (CCCHs) share a given time slot. Such a combined configuration is described in Chapter 7. This parameter is not transmitted but is derived from another parameter, CCCH_CONF.

BSC [GSM 03.02] Base station controller. Details are presented in Chapter 3.

BSIC [GSM 03.03] Base station identity code. An identifier for a BTS, although the BSIC does not uniquely identify a single BTS, since it has to be reused several times per PLMN. The purpose of the BSIC is to allow the mobile station to identify and distinguish among neighbor cells, even when neighbor cells use the same BCCH frequency. Because the BSIC is broadcast within the synchronization channel (SCH) of a BTS, the mobile station does not even have to establish a connection to a BTS to retrieve the BSIC. Figure G.9 shows the format of the BSIC. It consists of the network color code (NCC), which identifies the PLMN, and the base station color code (BCC).

BSN [Q.700–Q.704] Backward sequence number (7 bits). Used to acknowledge to the sender of a message (MSUs in SS7) that certain messages have been received. Chapter 8 presents more details.

BSS [GSM 03.02] Base station subsystem. Used to address all network elements belonging to the radio part of a GSM system. Parts of the BSS are the base transceiver station (BTS), the base station controller (BSC), and the transcoding rate and adaptation unit (TRAU). For more details, see Chapter 3.

BSSAP See *BSSMAP*.

BSSMAP [GSM 08.06, 08.08] Base station subsystem mobile application part. Both the BSSMAP and the direct transfer application part (DTAP) are

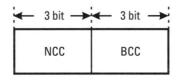

Figure G.9 Format of the BSIC.

users of the SCCP protocol of SS7 on the A-interface. BSSMAP and DTAP together form the BSSAP. The difference between the two is as follows:

- The BSSMAP is responsible for transmitting messages that the BSC has to process. This applies generally to all messages to and from the MSC where the MSC participates in radio resource management, for example, handover. The BSSMAP contains, furthermore, all messages for the administration of the A-interface itself.
- In contrast, the DTAP transports messages between the MS and the MSC, in which the BSC has just the relaying function, that is, it is transparent for these messages. These are all messages dealing with mobility management (MM) and call control (CC).

BTS [GSM 03.02]   Base transceiver station. Described in more detail in Chapter 3.

Burst [GSM 05.01, 05.02]   The nature of TDMA transmission is that radio energy is emitted in a pulsed manner rather than continuously. Mobile stations and BTSs send bursts periodically. Figure G.10 illustrates this for a GSM system in a power-over-time presentation. The actual data transmission is happening during the time period represented in Figure G.10 as a horizontal line. This time period is 148 bits, or 542.8 μs, long. Because GMSK—at least in theory—does not contain an amplitude modulated signal, the effective transmission power is constant over the entire transmission period. Figure G.10 also shows the specified corridor for the allowed power level of the signal over time. In total, a burst has a window of 577 μs, or 156.25 bit, before the next time slot starts. Physically speaking, the power level has to be reduced by 70 dB after 577 μs. These restrictions apply to the uplink as well as the downlink and determine the maximum number of bits an MS can send or receive at one time. The net bit rate is only 114 bits per burst, not 156.25. This reduced number of bits results from the mapping of a physical burst to a logical burst. The physical burst needs bits for administrative purposes that reduce the space available for signaling or user data. Note that all burst types specified for GSM follow a similar pattern:

- Each burst always begins with tail bits, which are necessary to synchronize the recipient. Tail bits are, except for the access burst, always coded as '000'.
- The tail bits are followed by 148 data bits, which differ in format for the various burst types.

- Each burst is terminated by another set of tail bits and the so-called guard period. This guard period is required for the sender to physically reduce the transmission power. The guard period is particularly long for the access burst, to allow mobile stations that are far from a BTS and hence experience propagation delays to also access the BTS (see *TA*).

The functional differences between the five logical bursts, defined for GSM, are as follows:

- Normal burst. The normal burst is used for almost every kind of data transmission on all channel types. The only exceptions to that rule are the initial channel request from the mobile station (CHAN_REQ/HND_ACC) sent in an access burst and the transmission of the synchronization data of a BTS that is done via the synchronization burst. All other data transfer on all traffic channels, dedicated control channels (DCCHs) and common control channels (CCHs) in uplink and downlink directions are done in normal bursts.

    Every normal burst contains 114 bits of useful data that are sent in two packets of 57 bits each. The so-called training sequence (TSC) is placed between the two packets. Note that the term *useful data* is not entirely accurate in this context, since the 114 bits are already channel coded and therefore contain some overhead (channel coding). Last but

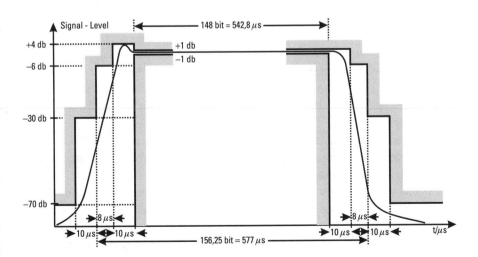

Figure G.10 The burst in the power-over-time presentation.

not least, there is a stealing flag between the training sequence and each data packet, which indicates to the recipient whether a 57-bit packet actually contains user data or FCCH information.
- Synchronization burst. The synchronization burst is used to transmit synchronization channel information (SCH) The synchronization burst uses a format similar to that of the normal burst (Figure G.12). In both cases, there are two data packets, left and right, from the training sequence. However, for the synchronization burst, each packet contains only a 39-bit payload, because the training sequence is 64 bits long. Note that the training sequence for the synchronization channel is identical for all BTSs and therefore allows a mobile station to easily distinguish an accessible GSM-BTS from any other radio system that accidentally works at the same frequency. Therefore, the training sequence in the synchronization channel serves two purposes: (1) It allows the mobile station to determine if there might have been transmission errors, and (2) it allows the mobile station to distinguish a GSM source from other transmission systems on the same frequency.
- Access burst. In contrast to the bursts described so far, the access burst comes in a rather unique format because of its special tasks (Figure G.12). A mobile station uses the access burst only for the initial access to a BTS, which applies in two cases: (1) for a connection setup starting from the idle state and (2) for handover (see under synchronized handover). In the first case, the MS sends the CHAN_REQ message in an access burst to the BTS. In the second case, the MS sends HND_ACC messages that also are mapped on access bursts.

In both cases the MS does not know the current distance to the BTS and, hence, the propagation delay for the signal (see *TA*). As long as the propagation delay is not known to the MS, the MS assumes it is zero. Therefore, it generally is uncertain if the access burst arrives within the receiver window of a BTS and how big the overlap is (Figure G.11). That is the reason for the lesser length of an access burst and the longer duration of the guard period. To ensure that an access burst arrives at the BTS during the proper time period the number of bits for the access burst was set to only 88 bits. The maximum distance between BTS and MS is, with this timing, about 35 km.

The normal burst would not fit into the receiver window if the unknown propagation delay was greater than zero. That is the reason why the normal burst is used only after the distance of the MS from the BTS is determined, and the MS is able to adjust its transmission accordingly. The adjustment parameter is called offset time and is

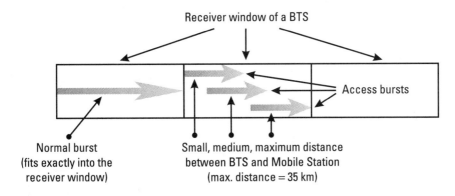

Figure G.11 The lesser length of an access burst.

calculated fairly simply. The BTS knows format and length of an access burst and is able to determine the actual propagation delay from when the signal arrives back at the BTS after being relayed by the MS. That also allows calculation of the distance of an MS from the BTS. The BTS provides the offset time to the MS, which in turn transmits its signal earlier, exactly by that time period (see *TA*).

The format of an access burst is also different from the other bursts. The access burst begins with 8 tail bits, rather than 3 as in the case of the other bursts, and the access burst always starts with the bit sequence 0011 1010$_{bin}$. The tail bits, together with the following 41-bit synchronization sequence which also always carries the same value, allows the BTS to distinguish the access burst from error signals or interfering signals. Hence, the access burst serves on the uplink a similar purpose as the synchronization burst does on the downlink. Nevertheless, in practice, it is common that the BTS determines background noise to be a CHAN_RQD message, as presented in Chapter 6.

The data field of an access burst is only 36 bits long and contains either a CHAN_RQD or an HND_ACC message. Note that both messages actually contain only 8 bits of "useful data."

- Frequency correction burst. The most simple format of all the bursts is used for the frequency correction burst, which is transmitted only in the frequency correction channel (FCCH) (Figure G.12). All 148 bits (142 bits + 6 tail bits) are coded with 0. A sequence of zeros at the input of a GMSK modulator produces, because of the peculiarities of the GMSK modulation, a constant transmitter frequency which is exactly 67.7 kHz above the BCCH median frequency. Therefore, the frequency of the FCCH is always 67.7 kHz above the frequency that is

Normal burst:

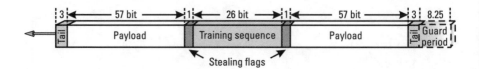

Synchronization burst:

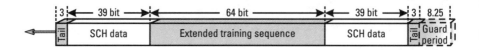

Access burst:

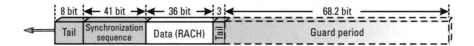

Frequency correction burst:

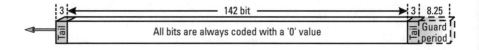

Dummy burst:

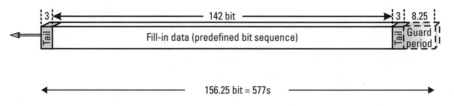

Figure G.12 The logical burst types.

advertised as the downlink frequency. This constant transmission frequency allows an MS to fine-tune its frequency to the BCCH frequency, to subsequently be able to read the data within the synchronization burst.
- Dummy burst. When the MS powers up, it checks the power level of the BCCH frequencies of the cells (BTSs) nearby to determine which BTS to use as a serving cell (Figure G.12). Similarly, when the MS is active, that is, involved in a call, the power level of the BCCH frequencies of the neighbor cells serve as basis for a possible handover decision.

To be useful as a reference, the BCCH frequency has to be transmitted with a constant power level. Thus, all time slots have to be occupied, and it is not allowed to apply power control on the downlink. For this purpose, the dummy burst was defined. These dummy bursts are inserted into otherwise empty time slots on the BCCH frequency. To prevent accidental confusion with frequency correction bursts, the dummy burst is coded with a pseudo-random bit sequence predefined by GSM.

C-Interface [GSM 09.02] The interface between the HLR and the MSC. More details are presented in Chapter 4.

Call reestablishment [GSM 04.08] A GSM functionality that currently is not used. Call reestablishment is applicable for speech connections only. In the case of an interruption of the radio connection during an ongoing conversation, the radio link failure procedure is invoked and the existing connection drops. With call reestablishment enabled, it is possible to prevent the disconnection of the conversation by establishing a connection to a suitable neighbor cell. Therefore, the mobile station will transmit a CHAN_REQ to this neighbor cell (cause: call reestablishment) that after SDCCH-assignment is followed by a CM_RES_REQ message (see Chapter 7) which is sent to the MSC for reconnecting the former MM and the CC connection. Since the GSM handover is a lengthy procedure with respect to time, it is possible that when the quality of the radio connection degrades rapidly the handover process could not be completed before the connection to that cell is lost completely. The call reestablishment functionality is addressing that behavior and allows a call to be maintained, even when the radio contact to the serving cell drops.

CB [GSM 03.41, 07.05] Cell broadcast. Synonymous with short message service cell broadcast (SMSCB). It allows a cell broadcast center (CBC) to send cell broadcast services (CBS), that is, text messages to all MSs in the entire

PLMN or parts thereof. Example applications of CB are traffic reports, weather forecasts, and stock quotes. In contrast to "regular" SMS, CB requires no confirmation from the mobile station. Another difference from SMS is that the CBC forwards broadcast messages directly to the BSC, bypassing the entire NSS. Figure G.13 illustrates this configuration. The BSC forwards the broadcast message over the Abis-interface to the BTS by facilitating an SMS_BC_REQ message or an SMS_BC_CMD message. The BTS, in turn, periodically transmits that information on the (cell broadcast channel (CBCH). When a BSS supports SMSCB, the CBCH is configured instead of the SDCCH/2.

A single CBS message may contain up to 82 bytes. In addition, it is possible to combine up to 15 CBS messages to form a so-called hypermessage.

CBC [GSM 03.41, 07.05]   Cell broadcast center. See *CB*.

CBCH [GSM 03.41, 07.05]   Cell broadcast channel. Used to transmit broadcast messages to mobile stations. The transmission rate of this optional channel is 782 bps. Network operators may choose to equip a CBCH instead of a SDCCH

CBS [GSM 03.41, 07.05]   Cell broadcast services. See *CB*.

CC   Call control. Application protocol between MS and MSC. See Chapter 7. Also, connection confirm message type of the SCCP (see Chapter 9.)

CCCH [GSM 05.01/05.02]   Common control channel. Generic term for all point-to-multipoint channels on the Air-interface. CCCHs are in the downlink

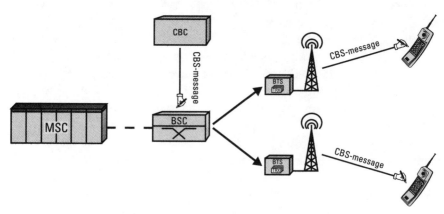

Figure G.13 Function of the CB/SMSCB.

direction, in particular the BCCH, the PCH, the CBCH, the AGCH. The only CCCH in the uplink direction is the random access channel (RACH). Network operators may configure the BCCH frequency to carry CCCHs in all even-numbered time slots (0, 2, 4, 6).

CCCH_CONF [GSM 05.02] Parameter of the BCCH/SYS_INFO 3 message that informs the MS about the actual configuration of the CCCHs in a cell. This information contains, in particular, whether a BTS uses a shared SDCCH/CCCH time slot (typically time slot 0) (BS_CCCH_SDCCH_COMB) and how many time slots are reserved for CCCHs (BS_CC_CHANS). Table G.5 lists all possible combinations.

CCITT Comité Consultatif International Télégraphique et Téléphonique. Organization that used to be responsible for international standardization of telecommunications-related issues. In 1993, CCITT was merged into International Telecommunication Union–Telecommunication Standardization Sector (ITU-T).

CDR Call drop rate. Indicator for the quality of service in a network. There is, however, no consistent understanding on how the CDR is to be determined. Some CDRs count dropped calls only if the drop occurs after the connection was already established; others also count unsuccessful call attempts.

For that reason, one has to be careful when comparing two systems based on the CDR. To properly determine the CDR, it is suggested to sum up all errors that are visible to the subscriber, that is, drops during outgoing handover, drops during mobile originating call setup, drops during mobile terminating call setup, and dropped calls during the stable phase of a call (without handover). Typically, the OMC measures the CDR by activating counters in the NSS or the BSS. (See Chapter 13.)

Table G.5
Relation Between CCCH_CONF, BS_CC_CHANS and BS_CCCH_SDCCH_COMB

CCCH_CONF	BS_CC_CHANS	BS_CCCH_SDCCH_COMB
0	1	No
1	1	Yes
2	2	No
4	3	No
6	4	No

CEPT Conférence Européenne des Postes et Télécommunications or Conference of European Postal and Telecommunications Administrations. Members are public administrations from European countries.

CFB, CFNRe, CFNRy, CFU, CUG, CW See *SS*.

CGI [GSM 03.03] Cell global identification. Identification composed of the location area identity (LAI) and the cell identity (CI). GSM allows use of both the CGI and the CI to identify a cell. Figure G.14 illustrates the format of the CGI.

Channel coding [GSM 05.03] Generic term that summarizes the different methods to protect data transmitted on the Air-interface from interference and transmission errors. In GSM, the fire code and the convolutional code are applied for that purpose. The devices performing channel coding are the mobile station and the TRX in the BTS. Channel coding adds information to the actual data, which allows the recipient to detect and sometimes even correct transmission errors. Figures G.15 and G.16 show the principles of channel coding for signaling and traffic channels (fullrate) respectively.

Figure G.16 shows that for channel coding the data in a TRAU frame (260 bits for fullrate) are separated into 182 class-1 bits (very important) and 78 class-2 bits (less important). Channel coding protects the two classes with different priorities. After channel coding is performed, the original data packet of 260 bits (user data) or 184 bits (signaling data) is extended to a length of 456 bits. The packet is then mapped on various bursts for the actual transmission. (see *Interleaving**).

Channel configuration Process of mapping signaling channels and traffic channels onto physical interfaces, for example, PCM 30. For the Air-interface, refer to Chapter 7; for the Abis-interface, Chapter 6; for the A-interface, Chapter 10.

Channel decoding [GSM 05.03] Reverse operation of channel coding. The received data are checked for errors; when errors are detected they are corrected

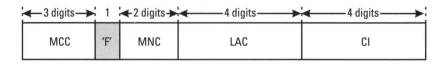

Figure G.14 Format of the CGI.

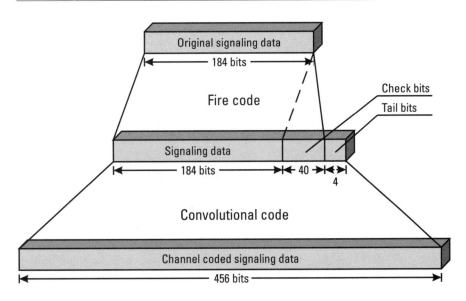

Figure G.15 Channel coding for signaling data.

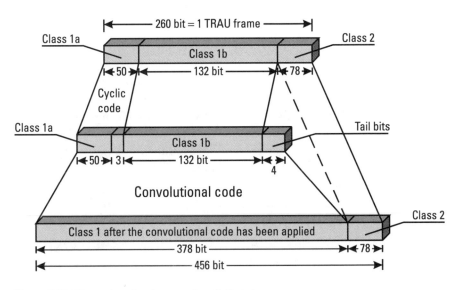

Figure G.16 Channel coding for user data (fullrate).

if possible. If the received data are found to be correct, the information that was added by channel coding has to be removed.

CI [GSM 03.03] Cell identity. A 2-byte-long hexadecimal identifier that, together with the location area (LAI) (see *CGI*), uniquely identifies a cell within a PLMN.

Ciphering [GSM 03.20] Used in GSM to encrypt data on the Air-interface between the mobile station and the BTS. Encryption applies only to the Air-interface. Therefore, tapping of a call still is possible on the terrestrial part of the connection. Precondition for ciphering is successful authentication. The process of authentication and activation of ciphering is performed in the following steps:

1. For each mobile station, the VLR stores up to five different authentication triplets (Figure G.17). Such a triplet consists of SRES, RAND, and K_c, and was originally calculated and provided by the HLR/AuC.

2. At first, the MS is sending a connection request to the network (e.g., LOC_UPD_REQ). Among others, this request contains the ciphering key sequence number (CKSN) and the mobile station classmark, which indicates what ciphering algorithms (A5/X) are available in the mobile station.

3. The NSS (more precisely, the VLR) examines the CKSN and decides whether authentication is necessary (see *CKSN*). Particularly to establish a second connection while another connection already exists (e.g., for a multiparty call), it is obvious that authentication is not required a second time during the same network access. A message is sent to the MS in case authentication is necessary. This DTAP message (AUTH_REQ) contains the random number, RAND, received from the HLR/AuC. The MS—more precisely, the SIM—uses the RAND and the value K_i as well as the algorithm A3 to calculate SRES (authentication procedure), as shown in Figure G.18.

4. The MS sends the result of this calculation, the SRES, to the VLR. The VLR compares the SRES that the MS has sent with the one that the HLR/AuC had sent earlier. The authentication is successful if both values are identical.

5. Immediately after calculating SRES, the MS uses RAND and K_i to calculate the ciphering key K_c via the algorithm A8.

6. To activate ciphering, the VLR sends the value K_c that the AuC has calculated and a reference to the chosen A5/X algorithm via the MSC and the BSC to the BTS.

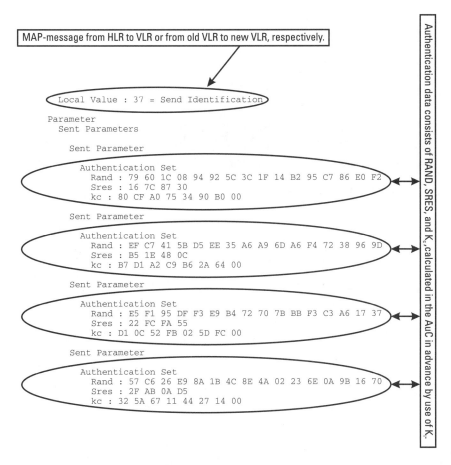

Figure G.17 Example of a (MAP) sendIdentification message, which, in this case, sends four authentication triplets for a mobile station to the VLR.

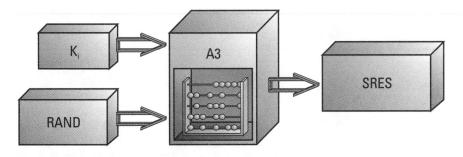

Figure G.18 Calculation of SRES from K_i and RAND by use of A3.

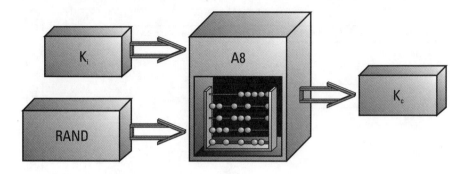

Figure G.19 Calculation of K_c from K_i and RAND by use of A8.

7. The BTS retrieves the cipher key K_c and the information about the required ciphering algorithm from the ENCR_CMD message and only forwards the information about the A5/X algorithm in a CIPH_MOD_CMD message to the MS. That message triggers the MS to enable ciphering of all outgoing data and deciphering of all incoming information. The MS confirms the change to ciphering mode by sending a CIPH_MOD_COM message. The ciphering process is illustrated in Figure G.20.

8. The algorithm A5/X uses the current value of the frame number (FN) at the time t_X together with the cipher key K_c as input parameters. The output of this operation are the so-called ciphering sequences, each 114 bits long, whereby one is needed for ciphering and the other one for deciphering.

9. As shown in Figure G.20, the first ciphering sequence and the 114 bits of "useful data" of a burst are XORed to provide the encrypted 114 bits that are actually sent over the Air-interface. Note that the ciphering sequences are altered with every frame number, which in turn changes the encryption with every frame number.

10. Deciphering takes place exactly the same way but in the opposite direction, as shown in Figure G.21.

Ciphering key sequence number   See *CKSN*.

CKSN [GSM 03.08, 03.20]   Ciphering key sequence number. A 3-bit-long value that references to a ciphering key, K_c. That is, when a particular K_c is stored in the MS and the MSC/VLR, a CKSN is assigned as well. The purpose is to allow the mobile station and the network a negotiation of the K_c to be

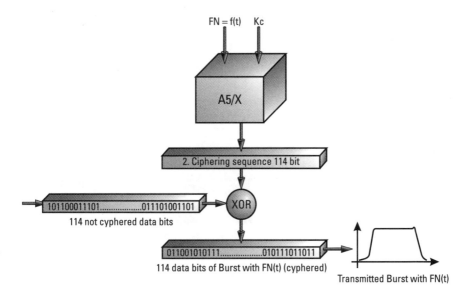

Figure G.20 Functionality of ciphering of data.

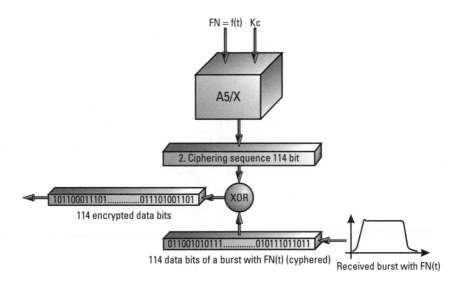

Figure G.21 Functionality of deciphering of data.

used, without compromising security by transmitting the value of K_c over the air. This applies particularly when an MS tries to establish an additional or subsequent operation with the network. In those cases, when the MS requests a connection, it sends its last valid CKSN as a parameter of the LOC_UPD_REQ- or CM_SERV_REQ-... message to the VLR. The VLR then decides, based on the CKSN, if ciphering can start immediately or if another authentication is required. The VLR may decide to request another authentication, even if the CKSN matches the VLR's entry.

CLIP, CLIR   See *SS*.

Closed user group   See *CUG*.

CM   Connection management.

COLP, COLR   See *SS*.

Comfort noise [GSM 06.12]   See *DTX*.

Constructor [X.208, X.209]   A data type of the TCAP protocol that is composed of primitives or other constructors. A primitive, in contrast to the data type constructor, is "atomic," that is, it contains only one parameter for TCAP. See Chapter 11.

Convolutional code   Procedure to secure data on the Air-interface against transmission errors. It is applied to signaling as well as to payload data. The idea is to add redundant information to the original data. The receiver then, by analyzing the redundant information, has the ability to detect errors and in some cases even correct corrupted data (channel coding).

The convolutional coder that GSM uses adds redundancy to the signaling data or to the payload by adding an additional bit to every input bit, thus doubling the amount of code. Hence, the code rate R equals one-half, that is, $R = 1/2$. The value of an added bit depends on the value of previous data bits stored in a shift register.

Figure G.22 provides the most simple convolutional coder, which is composed of a shift register, an element for modulus 2 addition, and a multiplexer. Note that this example is a simplification to illustrate the process. GSM uses a different procedure.

The length of the shift register determines the memory, which generally is referred to as *M*. The influence length (k) of such a coder indicates how many input bits are used to generate an output bit. In Figure G.22, k equals 2 ($k = 2 = M + 1$). A generator polynomial or a set of generator polynomials defines how a shift register and the input signal are coupled into the modulus 2 addition. That allows complete description of a convolutional coder.

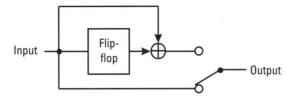

Figure G.22 Example of a convolutional coder.

Table G.6 is an example of coding where the flip-flop is preset with 0.

A so-called trellis diagram is a tool, or form of presentation, to determine the output code from the input data. This way of presentation is very helpful to better understand the principle of the decoder. The same example as above is used. Figure G.23 shows the input and output values, starting with the state s_0, where the flip-flop is preset with a zero value. The state and the transition between states are important. The value of both states, s_0 and s_1, is zero, the resulting code of this transition is 00. Note that the value of the output code and the value of the states (all 0) are only so for this case. The transition from s_2 to s_3, where the value changes from 0 to 1, is represented by the output code 11.

Convolutional coding generally allows for error detection and error correction, without the need for retransmission (forward error correction). The gain that is attained from the coding is evaluated by the bit error rate and depends largely on the applied method for modulation, demodulation, and decoding. Various models to simulate interference of a channel on the Air-interface were used to determine the bit error rate and find a suitable decoder. GSM's choice was the Viterby decoder.

CRC Cyclic redundancy check. A term from the transmission technique.

CUG [GSM 02.85] Closed user group. Used in telecommunications to establish groups of users with specific relationships and privileges. Originally, the concept was used in data networks for security reasons to protect access to a network from unauthorized access. The concept has been extended as follows.

Table G.6
Example of Coding Where Flip-Flop is Preset with 0

Transition	s_0–s_1	s_1–s_2	s_2–s_3	s_3–s_4	s_4–s_5
Input data	0	0	1	1	0
Output code	00	00	11	10	01

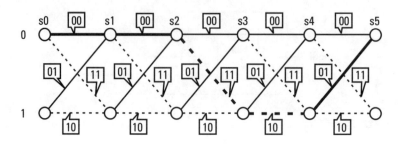

Figure G.23 Trellis diagram.

A CUG is a subset of subscribers of a PLMN (Figure G.24). Typical users of CUGs are companies (e.g., a shipping company) that have employees on the road and want to allow their employees to access the company resources but not have unlimited access to the rest of the network. This creates a kind of virtual private exchange or network. Basically, subscribers that belong to a CUG can communicate only with subscribers of the same group. That applies to calls both to and from a CUG member. The GSM supplementary service closed user group in its basic form restricts users from making any calls outside the CUG and does not allow a user to receive a call from someone outside the CUG. CUG has several options that grant special rights to individual users, like incoming access from non-CUG users or outgoing access to non-CUG users.

Each CUG has an identifier, and a subscriber can be a member of as many as nine different CUGs.

D1, D2   The two GSM 900 networks in Germany. The NDC for D1 is 171, the one for D2 is 172. This information is used in examples throughout this book.

D-interface [GSM 09.02]   The interface between VLR and HLR. See Chapter 4.

DCCH [GSM 05.01, 05.02]   Dedicated control channel. Generic term to address all bidirectional point-to-point control channels on the Air-interface. An example is the SDCCH.

DCS 1800   Digital Communication System 1800. A GSM system that was ported from the 900 MHz band to the 1800 MHz band. The DCS 1800 has more channels (374), but the protocol and the services are practically identical; only some minor changes to the protocol were made.

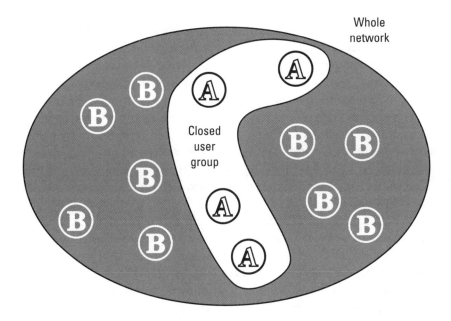

Figure G.24 Closed user group relative to the whole PLMN.

Digit 1 digit = 4 bits = 1/2 byte (see *BCD*).

div A mathematical operator for whole-number division. The result l of a *div* operation is always the whole-number part of the division. n by t, that is, the fraction of the division is discarded. The equation reads:

$$\lambda = v \ div \ \tau = v/\tau$$

where $\lambda, v \in N_0; \tau \in N$.

Examples are 0 *div* 25 F"Symbol"= 0, 1 *div* 2 = 0, 5 *div* 3 = 1, 10 *div* 3 = 3. A related operation is the *mod* operator.

Diversity Diversity reception is a concept in which the whole receiver path of the TRX module, including the antenna, is implemented multiple times, typically doubled. That allows the signal to reach the receiver via two different paths, and the receiver selects the better of the two signals. In practice, two antennas are mounted rather close to each other. When looking at a radio tower it typically is not possible to tell whether one or two antennas are mounted.

The reason for diversity becomes obvious when we think about an example from everyday life. While a driver is waiting at a stop light and listening to the radio, it frequently happens that the reception quality of the signal degrades. Often it is sufficient to move the car (i.e., the antenna) a very short distance to improve the quality of the reception.

A bad receiver signal at certain spots has the same cause for both radio broadcast and mobile radio, namely, same-channel interference. It is caused (partly) by extinction of the high-frequency signal caused by reflections (e.g., from buildings). This lies in the nature of (electromagnetic) waves, in which two signals with a propagation delay of l/2 extinct each other (l = wavelength). For a 900 MHz signal, l/2 equals about 1.5m.

With diversity, the TRX demodulates the two signals and forwards them to the low-frequency part. The low-frequency part then decides, based on the quality of the signal, which of the two to process further. Diversity is optional in GSM and hence a decision of the network operator whether or not to deploy it. Cost reduction is the most important factor for a network operator in the decision not to opt for diversity.

DLCI [GSM 08.06, 08.08]   Data link connection identifier. An identifier of a DTAP message on the A-interface that identifies which service access point identifier (SAPI) shall be used on the Air-interface. See Chapter 10.

DLR [ITU Q.711–Q.714]   Destination local reference. See Chapter 9.

Downlink   A direction of signal flow. Used for signals from the network to the mobile station. Table G.7 gives an overview of the frequencies and channels used for uplink and downlink on the Air-interface of GSM 900, DCS 1800, and PCS 1900.

Some time after GSM was put into service, the extended band was introduced to enlarge scarce frequency resources. The extended band was assigned exactly below the original GSM band. The additional channels are numbered from 975 to 1023, which avoids collision with already assigned channel numbers. A number of mobile stations do not support the extended band.

DPC [ITU Q.700–Q.704]   Destination point code. Identifier for the destination of an SS7 message (see *MSU*). The length of this identifier differs between ANSI and ITU. The ITU SS7 standard defines the DPC as 14 bits, while the ANSI SS7 standard defines it as 16 bits long. See Chapter 8.

Table G.7
Uplink and Downlink Frequencies of GSM 900, DCS 1800, and PCS 1900

	GSM 900	DCS 1800	PCS 1900
Downlink frequencies	935,2–960 MHz 925,4–935,0 Mhz (extended band)	1805–1880 MHz -/-	1930–1989,6 MHz -/-
Uplink frequencies	890,2–915 MHz 880,4–890,0 Mhz (extended band)	1710–1785 MHz -/-	1850–1909,6 MHz -/-
Channel numbers and number of available channels	1–124 = 124 channels 975–1023 = 49 channels (extended band)	512–885 = 374 channels -/-	512–810 = 299 channels -/-
Formula to convert frequency and channel number	n = ARFCN = 1–124	n = ARFCN = 512–885	n = ARFCN = 512–810
Downlink:	F(DL) = (935,2 + 0,2* (n − 1)) MHz	F(DL) = (1805,2 + 0,2* (n − 512)) Mhz	F(DL) = (1930 + 0,2* (n − 512)) Mhz
Uplink:	F(UL) = (890,2 + 0,2* (n − 1)) MHz	F(UL) = (1710,2 + 0,2* (n − 512)) MHz	F(UL) = (1850 + 0,2* (n − 512)) MHz
Extended Band:	n = ARFCN = 975–1023	-/-	-/-
Downlink:	F(DL) = (935,2 + 0,2* (n − 1024)) MHz	-/-	-/-
Uplink:	F(UL) = (890,2 + 0,2* (n − 1024)) MHz	-/-	-/-

DRX [GSM 03.13/05.02] Discontinuous reception. Used like the DTX as a power saver for the mobile station and to save radio resources. By separating mobile stations into paging groups, a particular mobile station needs to listen to the paging channels (PCH) only in certain multiframes. The transmitter can be switched off in the meantime, in what constitutes the power saving.

DTAP [GSM 04.08, 08.06]   See *BSSMAP*.

DTMF [GSM 03.14, 04.08]   Dual tone multifrequency. Method to transmit information in an in-band manner over telephone lines by facilitating a combination of two frequencies for every symbol. It distinguishes 12 different symbols. The symbols are the ones found on a telephone set, that is, the numbers 0 through 9 and the symbols * and #. It allows control of processes at the remote end of the connection. As shown in Figure G.25, each key is assigned a unique combination of two frequencies that are created in the telephone set and transmitted when the key is pressed. Today, DTMF is used worldwide for voice mail control, telephone banking, computer-integrated telephony applications, and so on.

GSM supports DTMF for voice connections only in the uplink direction. For GSM, however, no tones are sent over the Air-interface, but messages indicate the beginning and the end of a DTMF tone. When a user presses a button, the ASCII value of the button is sent to the MSC in a START_DTMF

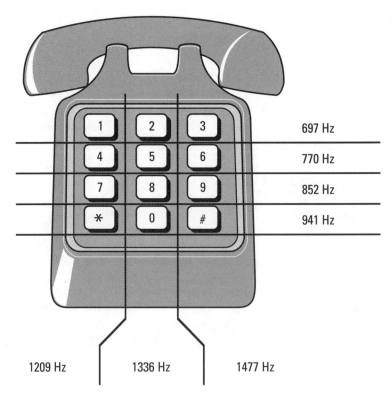

Figure G.25 Key/frequency combinations for DTMF.

Glossary

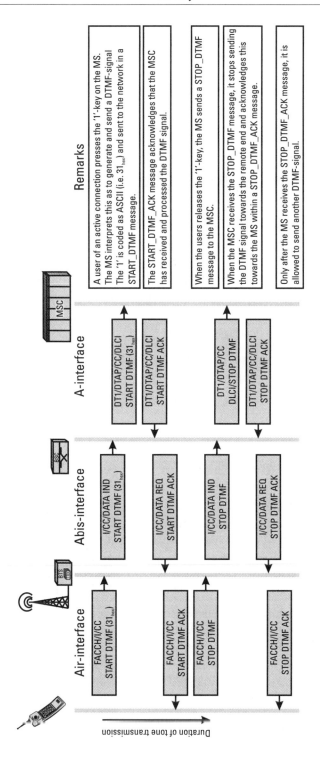

Figure G.26 DTMT-Transmission.

```
DATA INDICATION
 Channel Number: Bm + ACCH, TN = 7
 Link Identifier
 Channel Type: main signaling channel (FACCH or SDCCH)
 NA=0: applicable for this message
 SAPI 0
 L3 Information (Hex): 03 75 2C 31
 START DTMF
  Keypad Facility = 1
```

Figure G.27 Example of a START_DTMF message.

message. When the button is released, a STOP_DTMF message is generated. The messages are sent in encrypted form on the SACCH or the FACCH. The MSC analyzes the message, generates the frequency combination, and sends the DTMF signal to the remote end. Because the duration of a tone also might be important, the whole process becomes slightly more complex, as shown in Figure G.26.

Figure G.27 is an example of a START_DTMF message in which a user wants to send a 1. The 1 can be found coded as 31_{hex} in the last byte of "L3 Information (Hex)."

DTX [GSM 05.08, 06.12, 06.31, 06.32, 08.60] Discontinuous transmission. During a telephone conversation, typically only one party speaks at a time. At times, no one speaks. It is practical to switch off the Air-interface partly or completely during those silent times until the conversation resumes. One problem to avoid is clipping, that is, the situation when beginnings and ends of words are cut off because the volume of the speech is below a threshold and considered to be silent time. Setting the volume threshold is difficult because different languages have different dynamics, and what appears to be good enough for English may be poor quality for spoken Chinese. The process of detecting silent time and cease transmission is called discontinuous transmission. DTX needs to be distinguished from DRX (discontinuous reception); both methods are independent of each other.

DTX can be activated separately for uplink and downlink. The advantage of using DTX in the uplink direction is the power savings potential within the MS and for both uplink and downlink to reduce interference. One potential problem of DTX is related to background noise. People are so used to it that if it is not there, they assume the connection has been lost, particularly in a mobile conversation. DTX eliminates background noise, so to avoid the impression of a lost connection, artificial noise (called comfort noise) is generated when DTX is active.

With DTX enabled, the BTS or the MS sends only one block of data (456 bits according to channel coding) every 480 ms, which, because of interleaving and depending on the channel type, is transmitted with a variable number of bursts. That allows both sides to still measure the quality of the connection and to adjust the comfort noise if necessary.

Figure G.28 provides an example. After sending speech frames (see under TRAU frame), the MS indicates by a SID frame (SID = silence descriptor) that DTX was activated. A SID frame is a regular speech frame

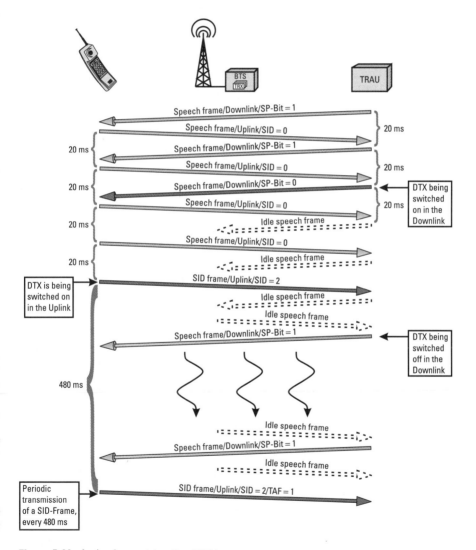

Figure G.28 Activation and details of DTX.

with a 320-bit length, where the SID control bits indicate that DTX was switched on and its data bits allow for the generation of the comfort noise. The TAF bit in the second SID frame from the MS indicates that this packet represents the compulsory frame that has to be sent every 480 ms.

The same applies for the activation of the DTX in the downlink direction. However, the downlink speech frame distinguishes between DTX on and off with the SP bit (speech indicator). This decision is taken by the TRAU.

Note that with active DTX, only the transmission on the Air-interface is turned off. Between TRAU and BTS, however, transmission of idle speech frames (see under TRAU frame) is required. DTX is switched off in the respective direction when one of the connected parties resumes the conversation.

The MSC is always in control of downlink DTX, while the uplink DTX has to be coordinated between BTS and MS. For that purpose, the BCCH/SYS_INFO 3 contains a 2-bit parameter, the DTX indicator, that indicates whether DTX can be applied to the uplink or if it needs to be prohibited.

E-interface [GSM 09.02, 09.10]   The interface between MSCs. See Chapter 4, Chapter 11, and Chapter 12.

EIR [GSM 03.02, 03.20, 12.02, 12.03]   Equipment identity register. See Chapter 4.

EMI   Electromagnetic interference. A critical factor, particularly in a system that uses pulses for transmission purposes. In a TDMA system, like GSM, the transmission occurs only a fraction of the time. Much research has been done in this area so far. The most important result was a potential temperature increase in parts of the head of a GSM-handset user. That is due to the high frequency radiation of a handset that is similar—but to a much lesser extent—to what happens in a microwave oven. This thermal aspect can be reduced by special antenna design, maximum distance between the antenna and the head, and low-power handsets (less than 2W). On the other hand, no certain evidence exists on the nonthermal impact of pulsed high-frequency radiation on the human organism.

It has been proven, however, that GSM handsets may affect electronic components of automobiles. For example, they possibly may activate the airbags, cause the ABS to fail, or cause the ignition system not to work properly. Therefore, it may be dangerous to use a mobile phone in a car without an external antenna. All manufacturers of handsets and all network operators nowadays advise subscribers to refrain from such use.

Another electromagnetic impact of a GSM phone is rather easy to reproduce. Anyone who has ever operated a GSM phone near other electronic instruments has noticed the low-frequency interference of an active GSM telephone with equipment like TVs, stereo receivers, even fixed network telephone sets. The reason why that can be heard at all lies in the TDMA structure of GSM. As shown in Figure G.29, the transmission energy is radiated in pulses every 4.615 ms. That corresponds to a low frequency of $f = 1/T = 217$ Hz. If that is not completely shielded—and in real life that is never the case—a part of the signal can be picked up by the electronic equipment.

Encryption See *Ciphering*.

ETSI European Telecommunications Standard Institute. European organization responsible for standardization in Europe. It emerged from CEPT in 1988.

Extended band See *Downlink*.

External handover [GSM 05.08] Handover between two cells assigned to two different BSCs. External handover consequently involves two BSCs and possibly two MSCs. (For more details, see Chapter 12; also see *T8* and *synchronized handover*.)

F-interface [GSM 09.02] The interface between the equipment identity register (EIR) and the MSC. More details are provided in Chapter 4.

FACCH [GSM 04.04, 05.01, 05.02] Fast associated control channel. An in-band signaling channel, just like the SACCH, that is associated with an active connection between the MS and the BTS. In contrast to the SACCH, which is sent once per multiframe, the FACCH is used only when no delay is acceptable, that is, if it is not possible to wait for the next SACCH. Then the FACCH is inserted instead of user data. The stealing flag serves to distinguish user data from signaling within a burst. The FACCH can transport 9200 bps in a fullrate channel and 4600 bps in a halfrate channel. (see *N201, Burst*.)

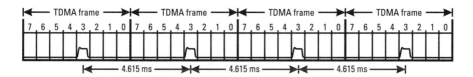

Figure G.29 Low-frequency interference caused by GSM.

FAS [G.711]   Frame alignment signal. Term from transmission systems. The FAS is used in a PCM system and transmitted in time slot 0. It allows for synchronization of sender and receiver on the frame structure.

FCCH [GSM 05.01, 05.02]   The BTS sends the frequency correction channel (FCCH) on time slot 0 of a BCCH-TRX in frequency bursts (see *Burst*). All 142 data bits are set to zero. Exactly five FCCHs are sent per 51-multiframe. The FCCH allows an MS to identify the frequency of a BTS in GSM. After sending an FCCH, an SCH has to be sent. More details are provided in Chapter 7.

FCS   Frame check sequence (FCS). Added to information and control fields of LAPD and SS7 frames. The task of the FCS is the cyclic redundancy check (CRC), which allows Layer 2 to detect transmission errors.

FDMA   Frequency division multiple access. Access-sharing technique like TDMA. Multiple access techniques are used to allow a number of users to access a system simultaneously. For that purpose, FDMA divides the frequency space into a multiplicity of frequencies that all can be used at the same time. GSM uses both types of access sharing. For more details, see Chapter 7.

FIB [ITU Q.700–Q.704]   Forward indicator bit. The most significant bit (MSB) within the second byte of an SS7 message (FISU, MSU, LSSU). It is used, together with FSN, BSN, and BIB, for error detection during data transmission. See Chapter 8.

Fire code   Part of a procedure for data protection during transfer over the Air-interface, used especially for signaling data. See *Channel coding*.

FISU [ITU Q.700–Q.704]   Fill-in signal unit. One of three SS7 messages of OSI Layer 2. (The other two are LSSU and MSU.) Figure G.30 shows the format of a FISU. The length indication (LI) element, which in the case of FISUs is always zero, allows distinguishing between FISU, LSSU, and MSU.

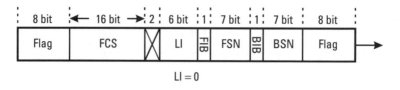

Figure G.30   Format of a FISU.

FN [GSM 05.01/05.02] Frame number. Internal clock of a BTS, to which every MS has to synchronize before the MS can start communicating with the BTS. For that purpose, the BTS broadcasts the current frame number five times for every 51-multiframe over the synchronization channel (see *SCH*).

The FN can take on values between 0 and 2,715,647, where each FN identifies exactly one TDMA frame within a hyperframe. The value 2,715,647 represents the possible number of frames, where $2{,}715{,}647 = (26 \cdot 51 \cdot 2048) - 1$. The -1 is necessary, since the count starts with zero. The equation represents the composition of a hyperframe. It consist of 2,048 superframes, each superframe consists of 26 multiframes with 51 TDMA frames or 51 multiframes with 26 TDMA frames. What is transmitted, however, is not the absolute value of the FN, but the relative position of an FN in the frame hierarchy, consisting of 51-multiframe, superframe, and hyperframe. (See also Chapter 7.)

This method of addressing the FN is similar to the way two people tell the time of day. Compare, for example, "The time is 54.900 seconds," and "The time is 3.15 p.m." In practice, the FN is sent as a combination of the parameters T1, T2, and T3, what could be brought in the analogy the example hours (T1), minutes (T2), and seconds (T3') of a clock. The rule is the following:

- T1 (11 bit): Number of the superframe in the hyperframe {0 ... 2,047}; T1 = FN div 1326, where $1{,}326 = 51 \times 26$
- T2 (5 bit): Number of the 51-multiframe in the superframe {0 ... 25}; T2 = FN mod 26
- T3: (6 bit): Number of the TDMA frame in the 51-multiframe {0 ... 50}; T3 = FN mod 51
- T3' (3 bit): (T3 − 1) div 10 out of {0 ... 4}

For T3, only the value of the decade has to be sent, since the synchronization channel is sent exactly five times per 51-multiframe, in fact, always in the position FN = 1, 11, 21, 31, 41 (compare Figure G.31). The single digit value, therefore, is redundant and there is no need for its transmission. The value T3' {0 ... 4} tells the MS exactly which FN in a 51-multiframe is meant and can easily calculate the FN of a 26-multiframe or the absolute value of FN.

Note that this rule applies only to the transmission of FN on the synchronization channel. When the CHAN_RQD message is being transmitted, the entire value of T3 has to be sent. That allows the number of the superframe (T1) to be truncated. Indeed, T1' is used in this case, rather than T1, with T1' = T1 mod 32. T1' represents the last five bits of T1. The reason for that is obvious:

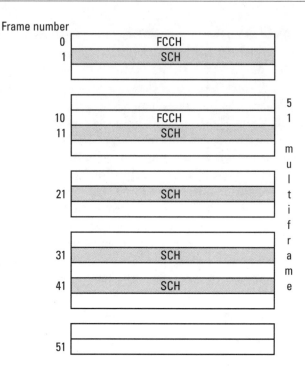

Figure G.31 The fixed position of a synchronization channel in a 51-multiframe.

- First of all, depending on the channel configuration, it is possible to send a RACH, practically anywhere within a 51-multiframe. Thus, T3 cannot be truncated.

- Furthermore, the BSC needs to respond to a CHAN_RQD within seconds. It is therefore not necessary to know the absolute number of the superframe. Knowing only the least significant five bits of the superframe number enables the BSS to uniquely identify and address a single CHAN_RQD message within a time period of $(2^5 - 1) \cdot 6.12s = 189.72s$ (a superframe has a cycle time of 6.12s), which is more than sufficient.

GSM refers to this type of frame number as starting time where *starting time* = FN mod 42432

Frame Number See *FN*.

Frequency hopping See *HSN*.

Frequency See *Downlink*.

FSN [ITU Q.700–Q.704] Forward sequence number. The sender of an MSU numbers each MSU sent (that is the FSN) and expects an acknowledgment from the receiver. The acknowledgment is sent in the form of the backward sequence number (BSN). See Chapter 8.

G-interface [GSM 09.02] The interface between VLRs. See Chapter 4 and Chapter 11.

G-MSC [GSM 03.02] Gateway mobile switching center. Mobile switching center with an additional functionality that allows a GSM network to interface with other networks. See Chapter 4.

Global title [Q.713] Optional part of the SCCP address. Various formats of the global title exist, but it always contains a routable number, which the SCCP uses to route messages to a network element. See Chapter 9.

GMSK [GSM 05.04] Gaussian minimum shift keying. Method for modulating signals in GSM. It is, as the name suggests, a special form of minimum shift keying (MSK), which belongs to the group of frequency modulation (FM) techniques. The modulated output signal F_O depends on the input signal E, where F_O is switched between the two frequencies $(F_T + f_t)$ and $(F_T - f_t)$. This represents the two (digital) input values $E = 0$ and $E = 1$. Figures G.32 illustrates the MSK modulation for a sequence of input bits. Figure G.32(a) shows the input signal that has to be modulated. The bit sequence in this example is 1110100110101000011. Figure G.32(b) shows the same bit sequence after two consecutive bits have been joined by an exclusive OR (XOR) operation.

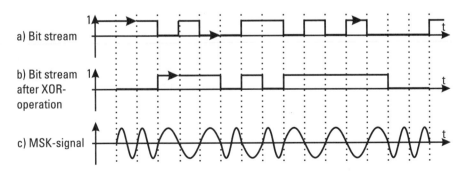

Figure G.32 Generation of an MSK-modulated signal

Table G.8 shows the corresponding truth table of this operation. To express it in words: When two consecutive bits have the same value, that is, both are 1 or both are 0, the result is 0; when two consecutive bits have different values, the result is 1. Figure G.32(c) and Table G.8 show how the output frequency of the sender depends on the result of the XOR operation.

A disadvantage of MSK is the resulting, relatively wide spectrum of this operation, due to the hard shift between the two frequencies $(F_T + f_t)$ and $(F_T - f_t)$. It is, however, crucial for every mobile system to use the scarce frequency resources as economically as possible. The GSM community, for that reason, decided not to use MSK; instead, it chose GMSK, which better meets the frequency economy constraint.

GMSK also uses the two frequencies $(F_T + f_t)$ and $(F_T - f_t)$ but shifts smoothly between the two. Figure G.33 illustrates the process. The bit sequence in Figure G.33(a) is the signal after application of the XOR operation. It represents, electrically speaking, a rectangular-shaped voltage. That voltage is then filtered by a low-pass filter, which smoothes the edges of the rectangle, as shown in Figure G.33(b). The new signal is used as input signal for the modulator. The resulting output frequency is shown in Figure G.33(c.) One can clearly see how the shift between $(F_T + f_t)$ and $(F_T - f_t)$ occurs more smoothly, which translates into a smaller frequency spectrum, that is, less bandwidth. This positive effect results from filtering the input signal with a Gauss filter with the following parameters

$$B \times T = 0.3$$

where B = 3-dB bandwidth and T = duration of an input bit: $T = 577\ \mu s / 156{,}25\ \text{bits} = 3{,}693\ \mu s$.

Table G.8
Truth Table for the Transmission Frequency

Bit ($N-1$)	Bit N	XOR	Frequency
0	0	0	$F_T + f_t$
0	1	1	$F_T - f_t$
1	0	1	$F_T - f_t$
1	1	0	$F_T + f_t$

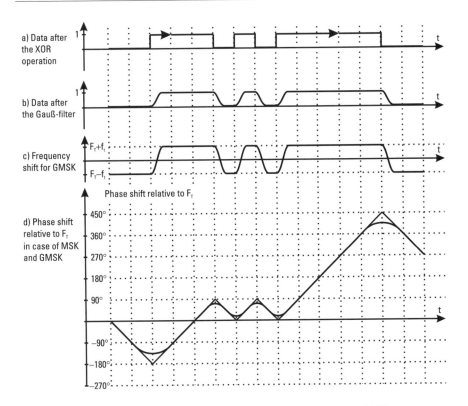

Figure G.33 Frequency chart of GMSK and phase chart of GMSK versus MSK.

From the available data and the index of the modulation $h = 0.5$, the frequency shift, f_t, can be derived. The value $\pm f_t$ indicates the extrema of the frequency, that is, the maximum and the minimum between which the carrier frequency is switched.

The following rule applies:

$$f_t = (data\ rate \cdot h)/2$$

The data rate is determined by the reciprocal value of T (the duration of 1 bit), $1/T = 270.8$ kHz.

$$f_t = 270.8 \text{ kHz} \cdot 0.5 \cdot 0.5$$

$$f_t = 67.7 \text{ kHz}$$

An interesting side effect is that since all 142 bits of the frequency correction burst (see *burst*) are coded with a zero value, the transmission frequency of a BTS is not exactly the BCCH frequency but is shifted by exactly 67.7 kHz upward.

The advantages of GMSK can be described as follows: (1) It does not—at least in theory—contain any AM portion, and (2) the required bandwidth of the transmission frequency is an acceptable 200 kHz.

H-interface [GSM 03.02]   The interface between the home location register (HLR) and the authentication center (AuC). The H-interface is not standardized, since the AuC is an integral part of the HLR.

Handoff   U.S. term for *handover*.

Handover [GSM 04.08, 05.08, 09.10]   Operation by which an MS is assigned another traffic channel while involved in a connection. It does not require the cell to change; the two channels can be on the same BTS. See Chapter 12.

Handover number [GSM 03.03, 09.10]   A number temporarily assigned to a subscriber in case of a handover between two MSCs, the so-called inter-MSC handover; used to route the call between the two MSCs. The format of this number corresponds to the MSRN.

Handover reference [GSM 04.08]   An 8-bit-long parameter that the destination BSC or new BSC randomly assigns for the handover process. The destination BSC sends the value both ways, to the destination BTS and via the new MSC to the originating MSC, which forwards the number, via originating BSC/BTS to the MS. The MS receives its handover reference in an HND_CMD message, which it transmits in an HND_ACC message to the destination BTS. The handover reference is therefore an identifier of the MS at the destination BSC or the new BTS.

HDLC   High-level data link control. The general frame format used, for example, by LAPD and SS7. An HDLC frame has, as shown in Figure G.34, a flag at each side, beginning and end, followed by an address field and a control field. The actual data follow the control field. The FCS allows detection of potential transmission errors.

Heading code [Q.704]   The message group and message type of an SS7 network management and test message are determined by the heading codes

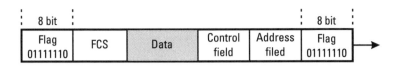

Figure G.34 The HDLC format

0 and 1. This group of messages, for example, COO (change over order) and COA (change over acknowledge), provides information used to bring links into service or to take them out of service, as well as to test and control connections. It is not used, however, to transfer user data.

HLR [GSM 03.02, GSM 03.20, GSM 09.02]   Home location register. See Chapter 4.

HLR number [GSM 03.03]   The format of an HLR number is the same as for a regular directory number, complying to ITU/T E.164 (see *MSISDN*). Every HLR is assigned a unique address, the HLR number, to enable the SCCP to route messages to the HLR (see also *global title*).

HMI [GSM 02.30]   Human-machine interface, formerly called the man-machine interface (MMI). Refers generally to the interface between the human user and any kind of machine (e.g., a personal computer). The keys and the display of the MS are the basic means of a GSM system for this interface, which defines which key combinations activate a certain feature.

HO   Handover (see also *Synchronized handover*). See Chapter 12.

HOLD   A supplementary service (see *SS*).

HSN [GSM 05.02]   Hopping sequence number. One of the following parameters that are necessary to execute frequency hopping.

- Cell allocation (CA). The list of all frequencies (see ARFCNs in ascending order) available in a cell (maximum 64) The CA is part of the BCCH / SYS_INFO 1.
- Mobile allocation (MA). Selection of the frequencies from the CA list that are applicable for the MS and the hopping sequence (maximum 64).
- Hopping sequence number (HSN). A value between 0 and 63 (6 bits without sign) used to control the hopping generator.

- Mobile allocation index offset (MAIO). A value between 0 and 63 (6 bits without sign). The valid range is always equal to the number of frequencies of the MA. The index is different for all MSs that occupy the same MA and the same time slot of a TDMA frame. This results in spreading the mobile stations over the available frequencies.
- Frame number (FN). The frame number or its partial counters (T1, T2, T3) are the variables that change over time for the generation of the hopping sequence.

The parameters MA, HSN, and MAIO are transmitted, for example, in the IMM_ASS-message.

Frequency hopping in a GSM system is referred to as slow frequency hopping because the frequency is constant, that is, the same for the duration of a burst. Fast frequency hopping, in contrast, requires altering of the frequency with every transmitted symbol (bit). The frequency in a GSM system, from the perspective of the mobile station, changes only from TDMA frame to the next TDMA frame, that is, every $8 \cdot 577\mu s$. The time for a mobile station to adapt to the new frequency, therefore, is $7 \cdot 577\mu s$. That applies if a synthesizer is available for both uplink and downlink. If only one synthesizer is available for both directions, the time to adjust to the new frequency is reduced to $\approx 4 \cdot 577\mu s$. The reason is that the mobile station can change only the frequency after the transmission in the uplink is completed (delay of three time slots; see *TA*).

The BTS, on the other hand, has to be capable of changing the frequency from burst to burst. To accomplish that, basically two alternatives are available.

- Baseband hopping. A TRX is divided into a baseband portion (for signal processing) and an HF portion, connected via a switch matrix. The switch matrix is able to connect every baseband signal with every HF signal and knows the hopping sequence of every channel. Uplink and downlink need to be switched separately, because of the known delay of three time slots between them. That results in the baseband signal portion of the TRX on the downlink being connected to a different carrier than for the uplink. The advantage of this alternative is that the HF portion is much easier to implement than for synthesizer hopping. The disadvantage is that it requires a switch matrix and it is possible to hop between only as many frequencies as currently are implemented.
- Synthesizer hopping. With synthesizer hopping, the frequency of the TRX changes from time slot to time slot by reprogramming the frequency of the synthesizer. That requires two pairs of synthesizers, one

pair for the uplink, the other for the downlink. One synthesizer of a pair is always active, while the other prepares for its new frequency, so that each has enough time to adjust to the new frequency. The guard period of a burst is actually used to switch between the synthesizers. Synthesizer hopping is, because of the strict requirements on frequency errors and phase faults in GSM, technically much more demanding than baseband hopping. One major advantage lies in the gained flexibility in network planning, where it is potentially possible to exploit all available frequencies for hopping, independent from the number of carriers that a BTS has.

Mixed configurations of baseband hopping and synthesizer hopping, separated for uplink and downlink, are viable and are deployed. A problem that exists, particularly with older versions of synthesizer hopping, is the potential interference with heart pacemakers (see *EMI*).

Hyperframe [GSM 05.01, 05.02] The hyperframe represents the largest time scale in the GSM frame hierarchy, with a total length of 3 hours, 28 minutes, 53 seconds, and 760 milliseconds. It is composed of 2,048 superframes, which are composed of 1,326 multiframes. See Chapter 7.

Idle-channel measurements [GSM 05.08, 08.58] The TRX permanently performs interference measurements on unused time slots, to determine potential interference on those channels. Interference can be caused by mobile stations or non-GSM systems. The measurements are sent to the BSC in an RF_RES_IND message. The BSC in turn takes the interference measurements into account before assigning a traffic channel.

IMEI [GSM 02.16, 03.03, 03.20] Mobile station equipment identity. Figure G.35 shows the format of the IMEI. In contrast to the IMSI, the IMEI identifies the mobile equipment rather than the subscriber. Another difference to the IMSI is that it is not mandatory for the network operator to query the

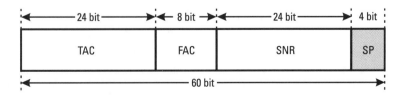

Figure G.35 Format of the IMEI.

IMEI. The purpose of the IMEI is to be a means for passive theft protection. When this functionality is active in a network, the EIR maintains information on stolen mobile equipment in a "black list," which makes stolen mobile equipment useless. It is even dangerous for a thief to use stolen equipment, since its use reveals the user's identity, which comes with the SIM, to the network operator. The IMEI comprises the following:

- A 24-bit-long type approval code (TAC). Before any mobile equipment can be brought into service, it has to undergo a test to show that it complies with safety regulations and functionality requirements. This process is called type approval, and the requirements are specified by GSM.
- An 8-bit-long final assembly code (FAC), which identifies the manufacturing facility.
- A 24-bit-long serial number.
- A spare field, currently not used.

Chapter 4 provides more details on the EIR.

IMEISV The IMEISV corresponds to the IMEI plus a software version number (SVN), which can be modified by the manufacturer in case of a software update. The format of the IMEISV is shown in Figure G.36.

IMSI [GSM 03.03, 03.20] International mobile subscriber identity. As identifier for a GSM subscriber, the IMSI is part of the subscriber data stored on the subscriber identity module (SIM) card. The IMSI uniquely identifies one subscription worldwide and is derived from ITU-T Recommendation E.212. Its structure is similar to the ISDN number, which is defined in ITU-T Recommendation E.164. The IMSI is a 15-digit number and is composed of the mobile country code (MCC), the mobile network code (MNC), and the mobile subscriber identification number (MSIN). Note that in GSM, unlike other standards, the MSIN of the IMSI is not used as the subscriber's telephone

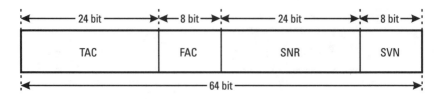

Figure G.36 Format of the IMEISV.

number. To make subscriber tracking more difficult, the IMSI is used only as an identifier when the temporary mobile subscriber identity (TMSI) is not available, e.g., for initial system connections. Figure G.37 shows the format of the IMSI.

IMSI attach/detach [GSM 04.08, 09.02] The BTS permanently broadcasts the parameter ATT in the BCCH / SYS_INFO 3 message. This parameter indicates whether the IMSI attach/detach procedure is required. IMSI detach is a procedure to inform the network that a mobile station will go into an inactive state and thus is no longer available for incoming calls, for example, because of powerdown or because the SIM is removed. The mobile station sends an IMSI_DET_IND message to the network each time it is powered down. The VLR keeps track of that state. The merit of this approach is that it saves radio resources and processing time. The call processing can switch to secondary call treatment, without the need of first sending a PAGING message and then waiting for expiration of the respective timers. Secondary call treatment means initiating call forwarding, voice mail, or simply indicating to the caller that the subscriber is currently not reachable. The complementary operation to IMSI detach is IMSI attach. It indicates to the network that a mobile station is active again. IMSI attach is related to periodic location updating. The location updating procedure is utilized to perform IMSI attach.

Inband signaling The counterpart to outband signaling. Inband signaling describes the situation when control information is transmitted within the traffic channel rather than in dedicated channels for signaling. Examples of outband signaling are SS7, LAPD signaling; examples of inband signaling are FACCH and SACCH (Air-interface).

Incoming call [GSM 04.08, 08.08, 09.02] A call request for a mobile subscriber. Also referred to as mobile terminating call (MTC). See Chapter 12.

Incoming handover [GSM 04.08, 05.08, 08.08, 09.02] During handover, the originating BTS or old BTS that the mobile station is leaving is involved in

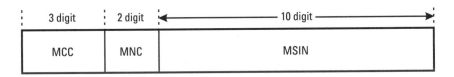

Figure G.37 Format of the IMSI.

an outgoing handover, while the destination BTS or new BTS to which the mobile station is handed over is involved in an incoming handover.

Interleaving [GSM 03.05, 03.50, 05.03]   Procedure to distribute or interlace the bits of a channel-coded block (see *channel coding*) onto several bursts. Since channel coding is designed to detect and correct errors on only a relatively few bits, it is the goal of interleaving to prevent complete loss of the information when a whole burst is corrupted. If, for example, a complete burst is lost, but all the others are transmitted without error, only one bit of a larger piece of information is missing and can be restored by the Viterby decoder.

The likelihood of group errors on a radio interface is naturally much higher than errors on single bits. The reason is the effect of fading, which typically is slower than the 270-Kbps transmission rate of the Air-interface.

For transmission of data, the bits are distributed even more than in the case of speech. For data transmission, it is even more important not to lose a single bit, since that could render a complete transmission useless. Speech is not very sensitive to single-bit errors. Propagation delay, on the other hand, is crucial for speech and does not have a very high priority for data connections. The more the bits of one sample are spread over time, the longer the receiver has to wait until all bits for a certain sample have arrived. For data services, that essentially affects only timers of the protocol. This affects the RLP protocol for nontransparent data and the end-to-end protocols of terminal applications for transparent data (GSM 03.05, 03.50).

In a fullrate speech channel, interleaving accounts for a maximum delay is 37.5 ms, while the maximum delay caused by the more intense interleaving in case of a fullrate data channel is 106.8 ms. Only RACH and SCH are transmitted without interleaving. Figure G.38 illustrates interleaving for a fullrate speech channel. The 456 channel-coded bits of block n are divided into 8 subblocks with 57 bits each and then rearranged. Subblocks 0 through 3 of block n are then interleaved with subblocks 4 through 7 of block $n-1$, while subblocks 4 through 7 of block n are interleaved with subblocks 0 through 3 of block $n+1$. Initially, subblocks 0 through 3 form the upper half of a burst, while subblocks 4 through 7 form the lower half of a burst. During the subsequent formation of the burst, the bits of the upper half alternatingly join with the bits of the lower half. Stealing flags are inserted in the middle of a burst.

Internal handover   A handover in which the BSC supports the handover procedure without support of the MSC. This is particularly the case for intra-BSC handover and intra-BTS handover. See Chapter 12; also see *T8, external handover*.

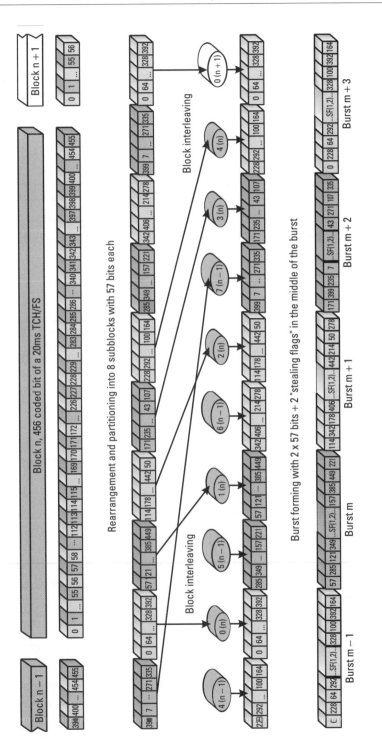

Figure G.38 Interleaving for a fullrate speech channel.

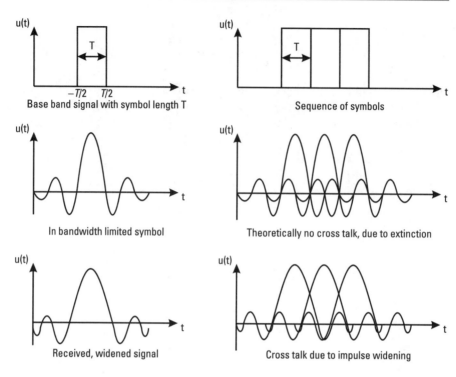

Figure G.39 Intersymbol interference.

Intersymbol interference Intersymbol interference refers to the cross-talk of neighbor cells due to widening of the impulses during transmission and caused by differences in the propagation delay. Figure G.39 illustrates this situation. The left side of the figure, from top to bottom, shows the transformation of a single symbol or impulse. The original signal is presented at the top, the signal after limiting the bandwidth is shown in the middle, and the widened signal after the transmission is shown below. As long as only one symbol is transmitted within the time frame ±T, no negative effect is experienced. The right side, however, shows what can happen if pulses are sent consecutively, one immediately after the other. Without widening, the pulses extinguish each other outside the time frame ± $T/2$. In case of pulse widening, cross-talk can occur and lead to errors during demodulation.

The reason for intersymbol interference during on-the-Air-interface is that the transmission channel is time variant, that is, it exposes time-dependent characteristics, which the demodulator has to take into account. Another source for intersymbol interference may stem from deviations between the clocks of the sender and the receiver of a signal.

ISDN Integrated Services Digital Network (ISDN).

ISUP [Q.761–Q.766] The ISDN user part. A user of OSI Layers 1 through 3, just like the SCCP. Although no GSM protocol, ISUP is used by the MSC for signaling purposes toward the ISDN.

ITU and ITU-T International Telecommunication Union–Telecommunication Standardization Sector.

IWF [PSTN/ISDN = GSM 09.03, 09.06, 09.07, CSPDN = GSM 09.04, PSPDN = GSM 09.05, 09.06] Interworking function. A subsystem in the PLMN that allows for nonspeech communication between GSM and other networks (see *PSTN, ISDN, CSPDN, PSPDN*). The tasks of an IWF are, in particular, to adapt transmission parameters and protocol conversion. The various IWFs typically are part of the MSC, where only the GMSC has to have IWFs. Physical manifestation of an IWF may be through a modem, which is activated by the MSC, dependent on the bearer service and the destination network. Table G.9 lists which modem types and transmission rates are available for data transmission between PSTN/ISDN and GSM. One can compare these transmission rates with the bearer services of GSM.

K_c **[GSM 03.08, 03.20]** The parameter K_c is the 8-byte-long cipher key and results from applying the two parameters K_i and RAND to the algorithm A8. Both the BTS and the mobile station use K_c for ciphering.

Table G.9
Valid Modem Types in GSM (According to GSM 09.07)

Network	Modem Type	Transmission Rate
PSTN	V.21	300 baud/synchron
PSTN	V.22	1200 baud/synchron + asynchron
PSTN	V.22bis	2400 baud/synchron
PSTN	V.23	1200/75 baud/asynchron
PSTN	V.26ter	2400 baud/synchron
PSTN	V.32	4800/9600 baud/synchron
ISDN	V.110	300–9600 baud/asynchron

K_i [GSM 03.08, 03.20]   The parameter K_i is an individual key that is different for every SIM card. Its value is a number of up to 16 bytes, which is used for authentication (see ciphering). K_i is known only to the SIM card and the AuC/HLR. Since data security and protection from misuse essentially depend on this information being secret, the value of K_i is part of the most secret data of the GSM world. That is the reason why K_i is never transmitted over any interface and is used only within the SIM card and the AuC/HLR (see *A3, A5/X, A8, K_c*).

LAC   See *Location area.*

LAI   See *Location area.*

LAPB [X.25]   Link access protocol B-channel. Described in ITU Recommendation X.25.

LAPD [Q.920, Q.921]   Link access protocol D-channel. See Chapter 6.

LI   Length indication.

Lm-Channel [GSM 04.03]   Lm-Channel is another name for the GSM halfrate channel with a transmission rate of 6.5 Kbps for speech.

LMSI [GSM 03.03]   Local mobile subscriber identity. A 4-byte-long parameter that the VLR assigns to a subscriber on a temporary basis. The intention is to expedite queries in the VLR. When the LMSI is assigned, both sides do not only use the IMSI but also the LMSI. Although there is no use for the LMSI in the HLR, except for queries to the VLR, it still must be stored in the HLR. Furthermore, it is required to send the LMSI whenever data between the two databases are exchanged.

Location area [GSM 03.03, 04.08]   A location area comprises at least one but typically several BTSs. A location area is defined for the following purpose:

- So a mobile station that changes the serving cell in the same location area does not need to perform a location update;
- So when the network tries to establish a connection to a mobile station, for example, for a mobile terminating call, it is necessary to send only the PAGING message to those BTSs that belong to the current location area of the MS.

Defining a location area, therefore, serves mainly one purpose, to reduce the signaling load. Every BTS broadcasts the location area via the parameter location area identity (LAI) in the messages BCCH/ SYS_INFOS 3 and 4. Even when the MS is involved in an active call, the location area still is communicated to the MS in the SYS_INFO 6 (this is particularly important in a handover). The format of the LAI is shown in Figure G.40. The shaded, one-digit field is a filler (1111_{bin}). It extends the only three-digit MCC to 2 bytes. That is not necessary for the two-digit (1-byte) MNC. The actual location area code (LAC) is four digits long. The LAC is an identifier that can be assigned by the network operator. All values, except 0000_{hex} and $FFFE_{hex}$, are allowed. Those two values are reserved for cases when the LAI on a SIM has been deleted.

LPD [GSM 04.06]   Link protocol discriminator. A 2-bit-long part of the address field of an $LAPD_m$ frame. See Chapter 7.

LSSU [ITU Q.700–Q.704]   Link status signal unit. One of three SS7 messages defined in OSI Layer 2 (the others are FISU and MSU). Figure G.41 shows the format of the LSSU. The LSSU is used to indicate error situations when a link is brought into or taken out of service. The figure also shows the possible values of the 3-bit-long status field. The distinction between LSSU, FISU, and MSU is achieved via the length indication (LI) element, which in an LSSU can take only the value 1 or 2. Although the value 2 is explicitly allowed by ITU, only the first byte—or rather the first 3 bits of the status field—are used to carry the actual status information.

LU   Location update. See Chapter 12.

MAIO   See *HSN*.

Mandatory   A term used in standardization to indicate that a specific parameter has to be included in a message. Typically, it also implies the general position that parameter has to be sent, namely, within the mandatory part that typically does not require a specific tag. The opposite term is *optional*.

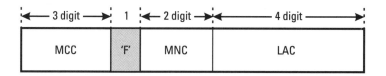

Figure G.40   Format of the LAI.

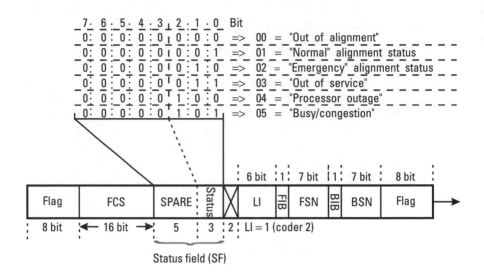

Figure G.41 Format of the LSSU.

MAP [GSM 09.02] Mobile application part. See Chapter 11.

MAX_RETRAN [GSM 05.08] One of the RACH control parameters sent in the BCCH/SYS_INFOS 1-4. MAX_RETRAN indicates how many consecutive times an MS may send the CHAN_REQ message to the BTS. Valid values are 1, 2, 4, and 7.

MCC [E.212] Mobile country code. A three-digit identifier that uniquely identifies a country (not a PLMN). Table G.10 is an extensive list of MCCs (see also *MNC, IMSI*).

MEAS_RES and MEAS_REP [GSM 04.08, 05.08, 08.58] BTS and MS measure the signal strength and quality of the received signal during an active connection. The MS periodically sends the measurements in a MEAS_REP message to the BTS (in a SDCCH/SACCH every 470.8 ms; on a TCH/SACCH every 480 ms). The BTS adds the measurements received from the MS to its own measurements and sends the result in a MEAS_RES message to the BSC. These measurements serve as input data for the BSC to perform the power control function and handover decision. Chapter 12 gives an example that shows how to decode a MEAS_RES message.

The measurements during an active connection should not be confused with the idle-channel measurements, which the TRX performs while time slots are not used. These results are periodically reported to the BSC in an RF_RES_IND message.

MM [GSM 04.08] Mobility management. A user protocol between the mobile station and the network switching subsystem (NSS), for which the base station subsystem (BSS) is transparent. The messages defined in MM allow for roaming and security functions (see *authentication*) in GSM.

MNC [GSM 03.03] Mobile network code. A two-digit identifier used (like the 3-bit-long NCC) to uniquely identify a PLMN (see also *MCC, IMSI, NCC*).

Table G.10
Mobile Country Codes

Country	MCC	Country	MCC
Afghanistan	412	Albania	276
Algeria	603	American Samoa	544
Angola	631	Antigua and Barbuda	344
Argentina	722	Australia	505
Austria	232	Azerbaijan	400
Bahamas	364	Bahrain	426
Bangladesh	470	Barbados	342
Belgium	206	Belize	702
Benin	616	Bermuda	350
Bolivia	736	Botswana	652
Brazil	724	British Virgin Islands	348
Brunei	528	Bulgaria	284
Burkina Faso	613	Burma	414
Burundi	642	Cambodia	456
Cameroon	624	Canada	302
Cape Verde	625	Cayman Islands	346
Central African Republic	623	Chad	622
Chile	730	China	460
Colombia	732	Comores	654

Table G.10 (continued)

Country	MCC	Country	MCC
Congo	629	Cook Islands	548
Costa Rica	712	Croatia	219
Cuba	368	Cyprus	280
Czech Republic	230	Denmark	238
Djibouti	638	Dominica (Commonwealth of the)	366
Dominican Republic	370	Ecuador	740
Egypt	602	El Salvador	706
Equatorial Guinea	627	Estonia	248
Ethiopia	636	Faroe Islands	288
Fiji-Islands	542	Finland	244
France	208	French Antilles	340
French Polynesia	547	Gabun	628
Gambia	607	Georgia	282
Germany	262	Ghana	620
Gibraltar	266	Greece	202
Greenland	290	Grenada	352
Guam	535	Guatemala	704
Guernsey (England)	234	Guiana (French)	742
Guinea	611	Guinea-Bissau	632
Guyana	738	Haiti	372
Honduras	708	Hong Kong	454
Hungary	216	Iceland	274
India	404	Indonesia	510
Iran	432	Iraq	418
Ireland	272	Isle of Man (England)	234
Israel	425	Italy	222
Ivory Coast	612	Jamaica	338
Japan	440 + 441	Jersey (England)	234
Jordan	416	Kenya	639
Kiribati	545	Korea (North)	467
Korea (South)	450	Kuwait	419
La Reunion	647	Laos	457
Latvia	247	Lebanon	415

Table G.10 (continued)

Country	MCC	Country	MCC
Lesotho	651	Liberia	618
Libya	606	Lithuania	246
Luxembourg	270	Macao	455
Madagascar	646	Malawi	650
Malaysia	502	Maldives	472
Mali	610	Malta	278
Mauritania	609	Mauritius	617
Mexico	334	Monaco	212
Mongolian Republic	428	Montserrat	354
Morocco	604	Mozambique	643
Namibia	649	Nauru	536
Nepal	429	Netherlands	204
Netherlands Antilles	362	New Caledonia	546
New Zealand	530	Nicaragua	710
Niger	614	Nigeria	621
Norway	242	Oman	422
Pakistan	410	Panama	714
Papua New Guinea	537	Paraguay	744
Peru	716	Philippines	515
Poland	260	Portugal	268
Puerto Rico	330	Qatar	427
Reunion	647	Romania	226
Russia	250	Rwandese Republic	635
San Marino	292	Sao Tome and Principe	626
Saudi Arabia	420	Senegal	608
Seychelles	633	Sierra Leone	619
Singapore	525	Slovakia	231
Slovenia	293	Solomon Islands	540
Somalia	637	South Africa	655
Spain	214	Sri Lanka	413
St. Kitts and Nevis	356	St. Lucia	358
St. Pierre and Miquelon	308	St. Vincent and the Grenadines	360
Sudan	634	Suriname	746

Table G.10 (continued)

Country	MCC	Country	MCC
Swaziland	653	Sweden	240
Switzerland	228	Syria	417
Taiwan	466	Tanzania	640
Thailand	520	Togo	615
Tonga	539	Trinidad and Tobago	374
Tunisia	605	Turkey	286
Turks and Caicos Islands	376	UAE	424
UAE/Abu Dhabi	430	UAE/Dubai	431
Uganda	641	Ukraine	255
United Kingdom	234 + 235	Uruguay	748
USA	310 – 316	Uzbekistan	250 + 434
Vanuatu	541	Venezuela	734
Vietnam	452	Virgin Islands	332
Wallis and Futuna Islands	543	Western Samoa	549
Yemen/Arab. Republic	421	Yemen/People's Republic	423
Yugoslavia	220	Zaire	630
Zambia	645	Zimbabwe	648

MOC Mobile originating call. See Chapter 12.

mod Mathematical operator that uses only the remnant of a division without fraction. The result λ of a (ν mod τ) operation is always an integer value of the division ν over τ. The equation reads:

$$\lambda = \text{mod } \tau = \nu - \tau \cdot (\nu \text{ div } \tau)$$

where $\lambda, \nu \in N_0, \tau \in N$. Examples are 3 mod 6 = 3; 10 mod 3 = 1; 10 mod 2 = 0 (see also div).

MoU Memorandum of Understanding. Worldwide forum for GSM network operators.

MPTY A supplementary service (see *SS*).

MS Mobile station. See Chapter 2.

MSC Mobile switching center. See Chapter 4.

MSC number [GSM 03.03] The format of the MSC number corresponds to that of an ordinary directory number (see *MSISDN*). Every MSC has its own routable number that allows the SCCP (see *global title*) to address messages to the MSC.

MSISDN [GSM 03.02, 03.12] Mobile subscriber ISDN. The directory number of a mobile subscriber. Note that it is possible for one subscriber to have several MSISDNs in parallel on one SIM. The different MSISDNs are used to address different services, for example, one number for voice, another number for fax. The format of the MSISDN is shown in Figure G.42.

Example: 49 171 5205787 is the directory number of a subscriber to the D1 network in Germany. The country code (CC) identifies a country or region (e.g., 49 for Germany, 1 for the United States); the national destination code (NDC) identifies the PLMN (e.g., 171 for the operator D1); and the subscriber number (SN) is a unique identifier within the PLMN.

MSK Minimum shift keying. A modulation technique (see *GMSK*).

MSRN [GSM 03.03] Mobile station roaming number. A temporary identifier, used for mobile terminating calls, to route a call from the gateway MSC to the serving MSC/VLR. The serving MSC/VLR is the MSC/VLR in which area the subscriber currently roams. The VLR assigns the MSRN when a request for routing information is received from the HLR. The MSRN is released after the call has been set up. Figure G.43 shows the format of an MSRN. The MSRN is used solely to route an incoming call and contains no information to identify the caller or the called party. See Chapter 12.

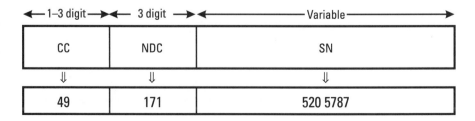

Figure G.42 Format of the MSISDN.

Figure G.43 Format of an MSRN.

The MSRN contains the following codes: the country code (CC) is the prefix of a country or region (e.g., 44 is the United Kingdom); the national destination code (NDC) identifies the PLMN (e.g., 172 is the D2 operator of Germany); and the temporary subscriber number (temp. SN) is assigned by the serving MSC/VLR of the called subscriber.

MS_TXPWR_MAX [GSM 05.08] Parameter that describes the maximum output power that an MS may use on a traffic channel of the serving cell.

MS_TXPWR_MAX_CCH [GSM 04.08, 05.05, 05.08] Parameter that describes the maximum output power that an MS may use on a random access channel (RACH). This parameter is sent to the MS in the BCCH/SYS_INFOS 3 and 4.

MSU [ITU Q.700–Q.704] Message signal unit. One of three SS7 messages defined in OSI Layer 2 (the others are FISU and LSSU). Figure G.44 shows the format of the MSU. The MSU is used for any kind of data transfer. The distinction between the various SS7 user parts (ISUP, SCCP, etc.) is made via the service information octet (SIO). The signaling data is carried in the signaling information field (SIF). The distinction between LSSU, FISU, and MSU is achieved via the length indication (LI) element, which in case of an MSU is always greater than 2. See Chapter 8.

MTC Mobile terminating call. See Chapter 12.

MTP [ITU Q.700–Q.704] Message transfer part. Part of SS7 that covers OSI Layers 1 through 3. Three messages are defined for the MTP: FISU, LSSU, and MSU. See Chapter 8.

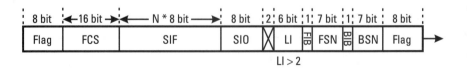

Figure G.44 Format of an MSU.

Multidrop A serial or ring configuration of BTSs on the Abis-interface. See Chapter 6.

Multiframe [GSM 05.01, 05.02] Two types of multiframes have to be distinguished in the frame hierarchy of GSM: the 26-multiframe, with a length of 120 ms, and the 51-multiframe, with a length of 235.8 ms. As the name suggests, the 26-multiframe is composed of 26 TDMA frames, while the 51-multiframe comprises 51 TDMA frames. Every TDMA frame comprises eight single bursts. In traffic channels, the 26-multiframe provides the absolute frame number (see *FN*), while the 51-multiframe provides the respective information in signaling channels.

Multiparty A supplementary service (see *SS*).

N200 [GSM 04.06 for LAPDm, Q.921 for LAPD] A counter for the maximum number of retransmissions on the Abis- or Air-interface. It defines how often one message can be retransmitted before the Layer 2 connection is torn down. For the Abis-interface, this value is 3 (N200 = 3). For the Air-interface, N200 can take on different values, depending on the channel type. For the Air-interface, it is necessary to distinguish between N200, which is the parameter for Layer 2 ,and RADIO_LINK_TIMEOUT, which is the respective tear down criterion of Layer 1. Table G.11 lists the values of N200 on the Air-interface.

N201 [GSM 04.06 for LAPD$_m$, Q.921 for LAPD] A counter for the maximum number of octets in the information field of the LAPD/LAPD$_m$ frames on the Abis- or Air-interface. The value of N201 on the Abis-interface is 260 (N201 = 260). The value for the Air-interface depends on the channel type. It has to be noted that the overall maximum length of a frame for signaling data is restricted, due to physical limitations (see *channel coding, GMSK*), to 23 octets or 184 bits. See Chapter 7. Table G.12 lists the values of N201 on the Air-interface.

N(R) and N(S) [Q.920, Q.921] Counters for forward error correction that indicate that a frame was received, N(R), or sent, N(S), respectively. Both counters are used by the LAPD and LAPD$_m$ protocol to acknowledge I-Frames (only) on the Air-interface and the Abis-interface. See Chapter 6.

NCC [E.164, GSM 03.03] Network color code. The 3-bit-long code that identifies the PLMN. This parameter is part of the BSIC and is broadcast in the synchronization channel (see *SCH, burst*).

Table G.11
Values of N200 on the Air-interface

Channel Type	N200
SACCH	5
SDCCH	23
FACCH (fullrate)	34
FACCH (halfrate)	29

Table G.12
Values of N201 on the Air-interface

Channel Type	N201
SACCH	18
SDCCH, FACCH	20
BCCH, PCH, AGCH	23

NDC [E.164, GSM 03.03] National destination code. Part of an ISDN number as defined by ITU-T in Recommendation E.164. Typically, the NDC addresses an area. It may, however, also be used to address a service, just as the NDC 800 addresses free phone service in the United States. In Germany, the NDCs 171 and 172 are used to address the two GSM 900 operators.

Neighbor cell All BTSs that are considered to be potential handover candidates for a given BTS. Each BTS (see *serving cell*) needs to inform all active and inactive mobile stations in its area about the BCCH frequencies of its neighbor cells. That is taken care of by the neighbor cell information within the BCCH/SYS_INFO 2 (for inactive mobiles) or BCCH/SYS_INFO 5 (for active mobiles).

Note that a handover can be performed into only those cells listed in the neighbor cell description of a BTS. If suitable or necessary neighbor cells are not listed, handover into such a cell is not possible, and the call drop rate (for drops during outgoing handover) will increase.

Nonsynchronized handover See *Synchronized handover.*

NSS Network switching subsystem. Generic term for the network elements in a PLMN that are not radio related. Hence, parts of the NSS are the mobile switching center (MSC), the home location register (HLR), the visitor location register (VLR), and the equipment identity register (EIR). See Chapter 4.

Ny1 [GSM 04.08] The maximum number of repetitions of a PHYS_INFO message during an asynchronous handover (see *T8, T3105*).

OACSU [GSM 04.08] Off-air call setup. Used to save radio resources by the late assignment of a traffic channel. In fact, no traffic channel is assigned before the called party takes the call. Without OACSU, the ASS_CMD message for the TCH assignment follows directly after CALL_PROC (for MOC) or after CALL_CONF (for MTC) (Figure G.45). Note that at this time no traffic channel is needed, because ringing or the creation of a ringback tone can be achieved without traffic channel.

With OACSU, the TCH gets assigned only when the called party actually takes the call and starts talking. Consequently, the scarce traffic channel resources experience, on average, a lower busy duration. This is particularly effective for short or unsuccessful calls. The drawback is that the user-to-user connection is not instantly available when the call is answered. Hence, no conversation is possible for a short period of time. GSM maintains that that time period is less than 1 second in 95% of all cases.

Octet 1 octet = 1 byte = 8 bit.

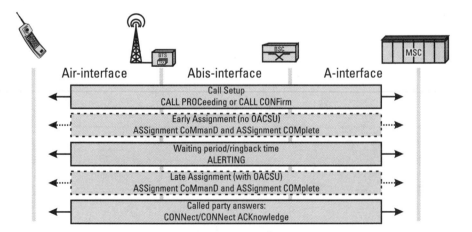

Figure G.45 Channel assignment with and without ASCSU.

OML [GSM 08.59]   Operation and maintenance link. See Chapter 6.

OPC [Q.700–Q.704]   Originating point code. Code that identifies the sender of an MSU. The ITU SS7 standard defines the OPC as a 14-bit code, while the ANSI SS7 standard defines it as a 16-bit code. The OPC is carried in the routing label, together with the DPC and the SLS. See Chapter 8.

OSI Reference Model   Open System Interconnection Reference Model. Defined by ITU in Recommendation X.200 as a means to facilitate communication between different systems. It contains general agreements on the communication among network elements. The basic underlying idea is to separate a communication process into seven largely independent layers. See Chapter 5.

OSS   Operation subsystem. Other name for the operation and maintenance center (OMC).

Outband signaling   See *inband signaling*.

Outgoing call   A call initiated by a mobile station, also referred to as a mobile originating call (MOC).

Outgoing handover   See *incoming handover, synchronized handover*, Chapter 12.

PABX   Private automatic branch exchange.

PAD [GSM 09.05]   Packet assembler/disassembler (Figure G.46). Required when a connection is routed from the PLMN to a PSPDN, either directly or via the PSTN.

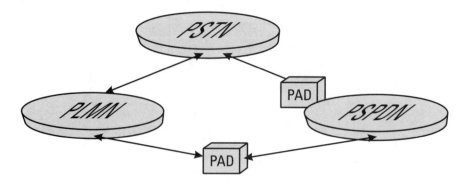

Figure G.46   Definition of the packet assembler/disassembler.

Paging group [GSM 05.02] Mobile subscribers are categorized into paging groups, to allow for discontinuous reception (see *DRX*), which extends the working time of a mobile's battery. Being assigned to a particular paging group, a mobile station needs to listen for an incoming call only every BS_PA_MFRMS 51 multiframes within a defined paging channel block (see Chapter 7). The IMSI and BS_CC_CHANS are necessary for the mobile station to calculate its paging group. BS_CC_CHANS indicates how many time slots of the BCCH frequency are reserved for common control channels (see *CCCH*).

The paging group itself (corresponding to the number of the paging channel block) is calculated as follows:

$$\textit{Paging group } (0 \ldots n-1) = [\textit{IMSI} \bmod 1000) \bmod (\text{BS\_CC\_CHANS} \times n)] \bmod n$$

where n = number paging channel blocks.

PCH [GSM 05.01, 05.02] Paging channel. The transmission rate per PCH block is 782 bps. See Chapter 7.

PCM Pulse code modulation.

PCS 1900 Personal Communication System 1900 (PCS 1900). PCS 1900 is not a standardized system but refers to a collection of mobile systems that operate in the 1900 MHz band in the United States. One of these systems is a derivative of GSM or DCS 1800. Other standards are CDMA/IS-95 and TDMA/IS-136.

PD [GSM 04.08] Protocol discriminator. See Chapter 7.

PDU [X.200–X.209] Protocol data unit. A term from the OSI Reference Model that refers to a data area in a message. The data area comprises application data and control information or coding information for the application data. PDU is used in the transaction capabilities application part (TCAP) as a generic term to address the dialog part and the component part or the content of these parts, respectively. For TCAP, the application protocol data unit (ADPU) is also used to address these areas. See *TCAP* and *MAP*.

PHS Personal Handy Phone System. An inexpensive Japanese standard for wireless communications in urban areas.

PIN [GSM 02.17] Personal identification number. A four- to eight-digit number that provides limited protection against unauthorized use. The PIN can be changed by the user and is stored on the SIM. The PIN is optional and can be disabled. When enabled, the PIN needs to be entered at powerup. When the wrong PIN is entered three consecutive times, the SIM goes into a blocked mode, and only the PIN unblocking key (PUK) can release the SIM. See Chapter 2.

PLMN Public land mobile network. A PLMN is the GSM, DCS, or PCS network of an operator within a country. The network color code (NCC) identifies a PLMN within a country.

PMR Private mobile radio. An inexpensive variant to provide mobile service. Several standards for PMR exist. Generally, PMR provides limited functionality and limited privacy. PMR uses simplex transmission, that is, only one party of a call can be on the air at a time.

Power class [GSM 05.05] In GSM, BTSs and mobile stations are classified and available in different power classes. The mobile station indicates its power class during every connection setup to the network. Most mobile stations belong to classes 4 and 5 (handheld). Mobile stations with higher output power typically are meant to be installed in cars or used for stationary applications.

Table G.13 lists the power classes for mobile stations and BTSs of GSM, DCS 1800, and PCS 1900.

Power control [GSM 05.05, 05.08] GSM requires that every mobile station is subject to power control. For the BTS, on the other hand, power control is optional. Depending on the quality of a connection, the BSC will request the BTS and the mobile station to adjust their output power. The purpose of power control is to minimize interference with other channels and to increase the working time of the battery.

The BSC informs the BTS via the Abis-interface within a BS_POWER_CON message of the output power to be used (Figure G.47). Only if necessary, the BSC will send an MS_POWER_CON message to the BTS to initiate an adjustment of the output power of the mobile station. This new output power level is forwarded to the mobile station within the Layer 1 header of the next SACCH to be sent. Note that one SACCH is sent to the mobile station every 480 ms, always telling the mobile station the current output power.

The maximum power is called P_n. Starting from there, the output power may be reduced in steps of 2 dB. Power control on the BTS side allows

Table G.13
Power Classes in GSM, DCS 1800 and PCS 1900

Class	GSM MS (W/dBm)	GSM BTS (W/dBm)	DCS1800 MS (W/dBm)	DCS1800 BTS (W/dBm)	PCS1900 MS (W/dBm)	PCS1900 BTS (W/dBm)
1	-/-	320/55	1/30	20/43	1/30	20/43
2	8/39	160/52	0,25/24	10/40	0,25/24	10/40
3	5/37	80/49	4/36	5/37	2/33	5/37
4	2/33	40/46	-/-	2,5/34	-/-	2,5/34
5	0,8/29	20/43	-/-	-/-	-/-	-/-
6	-/-	10/40	-/-	-/-	-/-	-/-
7	-/-	5/37	-/-	-/-	-/-	-/-
8	-/-	2,5/34	-/-	-/-	-/-	-/-
Micro (M1)	-/-	0,25/24	-/-	1,6/32	-/-	0,5/27
Micro (M2)	-/-	0,08/19	-/-	0,5/27	-/-	0,16/22
Micro (M3)	-/-	0,03/14	-/-	0,16/22	-/-	0,05/17

```
MS POWER CONTROL
    Channel Number: Bm + ACCH, TN = 2

    MS Power: 05h = +33 dBm

BS POWER CONTROL
    Channel Number: Bm + ACCH, TN = 3

    BS Power: 03h = Pn - 6 dB
```

Figure G.47 MS_POWER_CON and BS_POWER_CON messages from an Abis trace file.

reduction of the output power by 30 dB in 15 steps, while the output power of the MS can be reduced between 20 dB and 30 dB, depending on the standard (GSM, DCS1800) and the power class of the MS. Figure G.47 is an example of the two messages that are used for power control. While the MS_POWER_CON message always uses an absolute value, the BS_POWER_CON message always uses a relative value ($P_n - X$).

Note that all downlink channels of the BCCH-TRX have to permanently use the maximum output power P_n, since the BCCH is serving as a beacon and reference for the neighbor cell measurements of the mobile stations. Figure G.48 shows an example of a TRX where a variety of output power levels are used on the different channels. The channels TS 1 and TS 6 of this example are not in use.

Table G.14 provides the coding and the related output power for mobile stations in GSM and DCS1800/PCS1900. There is also a minimum output power below which a mobile station may not transmit. That value is 5 dB_m for GSM and 0 dB_m for DCS 1800/PCS 1900.

Presynchronized handover See *synchronized handover*.

Preprocessing [GSM 08.58] An optional functionality whereby the BTS evaluates its own measurement results and those from the mobile station. In this case, the BTS forwards a so-called PREPRO_MEAS_RES to the BSC only once in a while rather than sending MEAS_RES messages every 480 ms. Obviously, preprocessing should relieve the BSC from this task.

Primitive All parameters of TCAP that are "atomic," that is, that cannot be further partitioned, are called primitives. See Chapter 11. Furthermore X.200 (definition of the OSI Reference Model) defines primitives as messages that are exchanged between vertical layers of the OSI Reference Model (Layer N and Layer ($N+1$). See Chapter 5.

Pseudo-synchronized handover See *synchronized handover*.

PSTN Public switched telephone network.

PUK [GSM 02.17] PIN unblocking key. A 10-digit code stored on the SIM, which in contrast to the PIN, cannot be altered by the user. The PUK unblocks a SIM that was blocked because someone entered a wrong PIN three consecutive times.

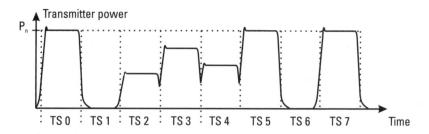

Figure G.48 Power control in the graph "Power over time."

Table G.14
Hex-Codes for the Output Power of an MS, as used in the MS_POWER_CON Message

Code	GSM	DCS 1800/PCS 1900
0	39	30
1	39	28
2	39	26
3	37	24
4	35	22
5	33	20
6	31	18
7	29	16
8	27	14
9	25	12
0A	23	10
0B	21	8
0C	19	6
0D	17	4
0E	15	2
0F	13	0
10	11	0
11	9	0
12	7	0
13	5	0
14	5	0
15	5	0
16	5	0
17	5	0
18	5	0
19	5	0
1A	5	0
1B	5	0
1C	5	0
1D	5	36
1E	5	34
1F	5	32

RACH [GSM 05.01, 05.02] Random access channel (RACH) is an uplink common control channel (CCCH) that the MS uses to send a connection request to the BTS. The access burst (see *burst*) is always used for the transmission of the RACH. The only two messages that are sent on the RACH are CHAN_REQ and HND_ACC, with a net data length of 8 bits and a transmission rate of 34 bps.

Radio_Link_Timeout   See *Radio link failure*.

Radio link failure [GSM 04.08, 05.08] Summary of the various conditions under which the connection over the Air-interface has to be regarded as terminated. Two possibilities need to be distinguished:

- Radio Link Failure, Layer 1. The most frequent reason for the declaration of a radio link failure on Layer 1 is simply the incapacity to decode the SACCH [RADIO_LINK_TIMEOUT] times. This definition applies to both uplink and downlink. For SACCHs that cannot be read, the channel decoder sets the BFI flag to 1. The parameter RADIO_LINK_TIMEOUT can be set per BTS and is broadcast in the BCCH / SYS_INFO 3. When a radio link failure on Layer 1 is detected on a dedicated channel (see *TCH* and *SDCCH*), the BTS sends a CONN_FAIL message with cause 1 = radio link failure to the BSC. The BSC consequently sends a CLR_REQ message cause 1 = radio interface failure to the MSC to indicate the connection loss and to release the resources. Figure G.49 shows this relation.
- Radio (data) link failure, Layer 2. The criteria for radio link failure on Layer 2 are similar to those for Layer 1. When an MS or a BTS sends a Layer 2 frame (I-frame, SABME-frame, etc.) as often as indicated in N200 without receiving a response, Layer 3 receives an error indication that requests release of that channel (Figure G.50). An ERR_IND message cause 1 = timer T200 expired (N200 + 1) times is sent to the BSC to release the channel. The BSC then performs the channel release procedure. This results frequently in a radio link failure on Layer 1, due to the fact that the radio link is so bad that the SACCH cannot be decoded anymore.

RAND [GSM 03.08, 03.20] Random number. As the name suggests, a number that the AuC picks on a random basis. The RAND range is up to $2^{128} - 1$ and has a length of 16 bytes. The network sends RAND in an AUTH_REQ message to the MS for authentication and ciphering.

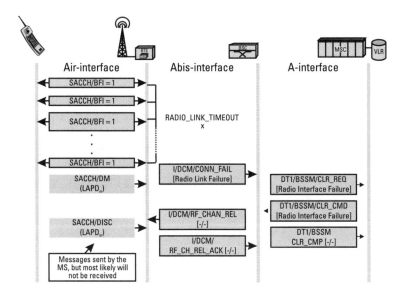

Figure G.49 A radio link failure, Layer 1.

RPE-LTP [GSM 06.10] Regular pulse excitation-long-term prediction. Procedure used by GSM to compress speech data from 64 Kbps to 13 Kbps in a fullrate channel and to 6.5 bps for a halfrate channel. This function is applied on both sides, the MS and the TRAU.

Roaming In the context of mobile communications, the ability of a subscriber to move around and change the location with or without an active connection. Wide-area roaming, perhaps internationally, distinguishes a mobile communications network like GSM from all wireless services.

Routing label [Q.700–Q.704] A 32-bit-long sequence within an MSU, consisting of the destination point code (DPC), originating point code (OPC), and signaling link selection (SLS) (Figure G.51). The routing label is used for message routing and is part of the signaling information field (SIF). See Chapter 8.

RSL Radio signaling link. Signaling channel on the Abis-interface between TRX and BSC. The signaling channel between BTS and BSC to convey information related to O&M is correspondingly referred to as OML.

RSZI [GSM 03.03] Regional subscription zone identity. Code that restricts the service area for a particular subscriber by restricting the roaming capability within a PLMN. The RSZI consists of a country code (CC), the network

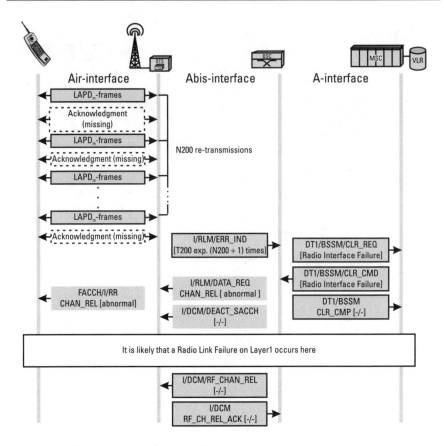

Figure G.50 Message flow when a radio link failure on Layer 2 occurs.

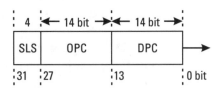

Figure G.51 Format of the routing label.

destination code (NDC), and a zone code (ZC). RSZIs can be assigned to a subscriber on a per-PLMN basis to indicate the allowed service areas of a PLMN for that subscriber. Figure G.52 is an example of an RSZI for a subscriber to the D2 network in Germany.

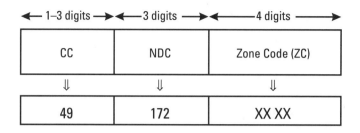

Figure G.52 Format of the RSZL (CC = 49, for Germany; 172 = D2 network).

RXLEV [GSM 05.08] RXLEV provides the results of the measurement of the receiving level on the Air-interface. These measurements are performed independently by the MS and the BTS. The BTS measures the receiving level for an active connection. The MS measures the receiving level of that BTS, where an active connection exists (serving cell) plus the receiving level of the neighbor cells indicated in the SYS_INFO 2. The values of both RXLEV and RXQUAL are sent to the BSC in a MEAS_RES/MEAS_REP, as basis for a decision by the BSC on power control or handover. It has to be distinguished between values with active DTX (SUBSET) and values without DTX (FULL or ALL). (See Chapter 12.) RXLEV values are coded the same way for uplink and downlink. A 5-bit-long, binary coded RXLEV value can directly be converted into a receiver-level dB_m. The procedure outlined in Table G.15 is applicable.

RX_LEV_ACC_MIN [GSM 05.08] Indication of the minimum receiving level that an MS has to receive from a BTS for that BTS to be a candidate as serving cell. Typical values are around −100 dB_m. Every BTS broadcasts this parameter in the BCCH / SYS_INFOS 3 and 4.

Table G.15
Conversion of RXLEV

RXLEV	Receiving Level (dB_m)
0	< −110
1–62	−(111 − RXLEV) through −(110 − RXLEV) (level range)
63	> −48

RXQUAL [GSM 05.08] RXQUAL values, like the ones for RXLEV, are relevant for the decision of a BSC on power control and handover. They indicate the bit error rate that was measured on the Air-interface. The bit error rate can be determined by facilitating the training sequence (see *TSC*). The current values of RXQUAL for an active connection can be tracked in the MEAS_RES/MEAS_REP messages by using a protocol analyzer on the Abis-interface. Values with active DTX (SUBSET) have to be distinguished from values without DTX (FULL or ALL).

RXLEV values are coded the same way for uplink and downlink and are 3 bits long, which allows for a range of 0 through 7. Coding of RX_QUAL is shown in Table G.16.

SACCH [GSM 04.04, 05.01, 05.02] Slow associated control channel (see also *FACCH*). The inband control channel assigned to the TCH or the SDCCH. Every 26th burst of a TCH or every 51st burst of an SDCCH is an SACCH. The consequence is that exactly one SACCH is sent per multiframe. Figure G.53 illustrates the format of the SACCH for uplink and downlink. The transmission rate is 391 bps when the SACCH is assigned to the SDCCH. When assigned to the TCH, the transmission rate is 383 bps.

SAPI [Q.920, Q.921, X.200] Service access point identifier. See Chapter 6.

SCH [GSM 04.08, 05.01, 05.02] Every BTS broadcasts the synchronization channel (SCH) in time slot zero of the BCCH-TRX. The SCH contains the absolute value of the frame number (see *FN*) of a BTS, which is time

Table G.16
Conversion of RXQUAL in Bit Error Rates

RXQUAL	Bit Error Rate	Assumed BER
0	BER < 0, 2 %	0, 14 %
1	0, 2 % < BER < 0, 4 %	0, 28 %
2	0, 4 % < BER < 0, 8 %	0, 57 %
3	0, 8 % < BER < 1, 6 %	1, 13 %
4	1, 6 % < BER < 3, 2 %	2, 26 %
5	3, 2 % < BER < 6, 4 %	4, 53 %
6	6, 4 % < BER < 12, 8 %	9, 05 %
7	BER > 12, 8 %	18, 1 %

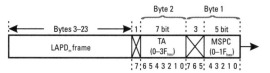

Payload: e.g., MEAS_REP message of Uplink; e.g., SYS_INFO 5 + 6 of Downlink
TA: Timing advance (as measured by the BTS)
MSPC: MS power control (the power level that the MS actually uses, or will use

Figure G.53 Format of the SACCH.

dependent, and the base station identity code (see *BSIC*) for an initial rough identification of the cell. The SCH has a length of 25 bits and is sent in the synchronization burst.

SCCP [Q.711–Q.714] Signaling connection control part. Defined in OSI Layers 3 and 4. The SCCP is a user of the SS7-MTP 3 and is used by GSM on almost all NSS interfaces, including the A-interface. See Chapter 9.

SDCCH/4 and SDCCH/8 [GSM 05.01, 05.02] Standalone dedicated control channel. Used for uplink and downlink of the Air-interface to transmit signaling data for connection setup and location update (LU). The transmission rate is 779 bps. The distinction between SDCCH/8 and SDCCH/4 refers to the channel configuration on the Air-interface. An SDCCH/8 channel configuration can never be realized on TS 0 of the BCCH-TRX. The available bandwidth is, because of the BCCH that occupies part of the bandwidth there, not sufficient to allow for that. TS 0 of the BCCH-TRX can be used for, at maximum, one SDCCH/4 channel configuration with four SDCCH subchannels. See Chapter 7.

Serving cell The BTS where the MS is actually booked on and from which the MS receives BCCH information.

SFH Slow frequency hopping (see *HSN*).

SI [Q.700–Q.704] Service indicator. A 4-bit-long parameter within an MSU that indicates the user part (SCCP, ISUP, etc.). See Chapter 8.

SID [GSM 06.31, 6.41] Silence descriptor. A 2-bit-long parameter that indicates whether a TRAU frame in the uplink is a SID frame (SID = 2) or a regular speech frame (SID = 0). A SID frame contains only data for the purpose of producing comfort noise (see *DTX*) but no speech information.

SIF [Q.700–Q.704]   Signaling information field (see also MSU). The maximum length of a SIF amounts to 272 bytes. See Chapter 8.

SIM [GSM 02.17, 11.11]   Subscriber identity module. See Chapter 2 (see *PIN*, *PUK*).

SIO [Q.700–Q.704]   Service information octet. Used in SS7 messages to distinguish between national and international messages, as well as between the various SS7 user parts (see *MSU*); 8 bits long. See Chapter 8.

Skip indicator   See *Transaction identifier*.

SLC [Q.700–Q.704]   Signaling link code. Part of the routing label of SS7 management and test messages. It is located at the same place as the SLS field of other SS7 messages and is 4 bits long. The SLC can be used to identify the SS7 connection when a management or test message is sent (SLS = 0, if nothing is defined). The SLC field is not used for load sharing.

SLR [Q.711–Q.714]   Source local reference. See Chapters 9 and 13.

SLS [ITU Q.700–Q.704]   Signaling link selection. Part of the routing label of SS7 Messages. This 4-bit-long parameter is used to balance load on SS7 connections. See Chapter 8.

SMG   When the standardization work on GSM started, it was carried out by a group called Groupe Spécial Mobile (GSM). When more and more countries worldwide became interested in GSM, it was decided to change the name of the group to Special Mobile Group (SMG). At the same time, the meaning of GSM was changed to Global System for Mobile Communications. SMG is the standards body responsible for all GSM specifications. SMG is a technical committee of ETSI.

SMS [GSM 03.40]   Short message services. A GSM service that allows small text messages to be sent to and from a mobile station. It is similar to a paging service integrated in the MS but is two-way.

SMS gateway MSC [GSM 03.02]   See *SMS interworking MSC*.

SMS interworking MSC [GSM 03.02]   The interface functionality for short messages from the SMS service center toward the mobile station is performed by a special MSC, called SMS gateway MSC. Routing messages in the opposite direction, from the mobile station to the SMS service center, is done by another MSC, called the SMS interworking MSC. Figure G.54 illustrates the tasks of the two types of MSCs.

SMSCB   Short message service cell broadcast. See *CB*.

SPC [Q.700–Q.704]   Signaling point code. An identifier in SS7 to specify the address of a network element, for example, that of a BSC. SPC is the generic term for originating point code (OPC) and destination point code (DPC). See Chapter 8.

SRES [GSM 03.08, 03.20]   Signed response. Used to verify the identity of an MS. SRES, which is 4 bytes long, is calculated by the MS by applying K_i and RAND to the algorithm A3 (see *Ciphering*).

SS [GSM 02.04, 04.10, 04.80]   Supplementary services. Telecommunications services that supplement or modify a basic service. This terminology is used by GSM and ISDN. Examples of supplementary services are call forwarding and multiparty call.

In addition to those SSs that GSM 02.04 lists, there exists another category of supplementary services, the unstructured supplementary services (USSs). USSs give mobile stations and the network the capability of exchanging PLMN-specific text messages up to 80 bytes long. The text messages are called unstructured supplementary services data (USSD). This allows the network operator to offer proprietary features to its subscribers. USSD can be created on the MS side, by means of the HMI/MMI keys.

Table G.17 lists and explains all current GSM supplementary services.

SS7 [Q.700–Q.704]   Signaling system number 7 (SS7). See Chapter 8.

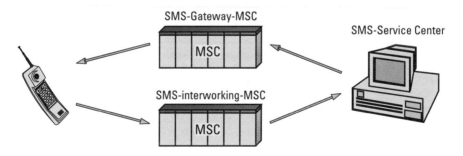

Figure G.54   The task of the SMS gateway MSC and the SMS interworking MSC.

Table G.17
GSM Supplementary Services

Acronym	GSM	SS-Code (Hex)	SS Name	Explanation
CLIP	02.81	11	Calling Line Identification Presentation	The number of the calling party is sent to the MS of the served subscriber for display.
CLIR	02.81	12	Calling Line Identification Restriction	The number of the served subscriber is restricted from being displayed at the called party's terminal.
COLP	02.81	13	Connected Line Identification Presentation	The number of the called party that the call is connected to is sent to the MS of the served subscriber for display.
COLR	02.81	14	Connected Line Identification Restriction	The actual number of the served subscriber, when answering a call, is restricted from being displayed at the calling party's terminal.
CFU	02.82	21	Call Forwarding Unconditional	A number to which all incoming calls will be redirected that has to be registered.
CFB	02.82	29	Call Forwarding on Mobile Subscriber Busy	Incoming calls will be redirected to a previously registered number when the MS is in a busy state.
CFNRy	02.82	2A	Call Forwarding on No Reply	Incoming calls will be redirected to a previously registered number when the MS does not answer within a specified time. The served subscriber may adjust the *No Reply Timer*.

Table G.17 (continued)

Acronym	GSM	SS-Code (Hex)	SS Name	Explanation
CW	02.83	41	Call Waiting	When the MS is in the busy state, Call Waiting indicates to the served subscriber that a second call came in.
HOLD	02.83	42	Call Hold	This allows the served subscriber to put a call on hold.
MPTY	02.84	51	Multi Party Service	MPTY allows the served subscriber to establish a conference call with up to 5 peer parties.
CUG	02.85	61	Closed User Group	CUG allows restriction of the communications capabilities to a limited number of parties. Several options exist to allow for incoming and outgoing access to parities outside the CUG. A subscriber can be member of up to ten Closed User Groups
AoCI	02.86	71	Advice of Charge (Information) (AoCI)	AoCI is the information level of AoC. It allows a MS to calculate a fairly accurate estimate of the call charges that will incur, by means of the *AoC parameters* that the MS receives at call set up. For MOC, these parameters are sent in a Facility element of a CON message, for the MTC (roaming charges) in a FACILITY message, which directly follows the CON message.
AoCC	02.86	72	Advice of Charge (Charging) (AoCC)	The same principle applies for the charging level of AoC, as for AoCI. The difference is the added requirement on accuracy. It is possible to limit the total amount of charges. When this limit is reached, an existing call is dropped and new calls are not accepted. AoCC can be used for pre-paid services.
BAOC	02.88	92	Barring of All Outgoing Calls	The subscriber is not able to place an outgoing call.
BOIC	02.88	93	Barring of Outgoing International Calls	The subscriber is not able to place outgoing international calls.

Table G.17 (continued)

Acronym	GSM	SS-Code (Hex)	SS Name	Explanation
BOIC-exHC	02.88	94	Barring of Outgoing International Calls except those directed to the Home PLMN Country	The subscriber is not able to place outgoing international calls, except to the country where the home PLMN is.
BAIC	02.88	9A	Barring of All Incoming Calls	The subscriber is not able to receive calls.
BIC-Roam	02.88	9B	Barring of Incoming Calls when Roaming Outside the Home PLMN Country	The subscriber is not able to receive calls when roaming outside the country where the home PLMN is (to prevent charges for international roaming).
USS	02.90	F0 - FF	Unstructured Supplementary Services	This allows a network operator to offer PLMN specific, i.e., non-standardized Supplementary Services.

SSF [Q.700–Q.704] Subservice field. A term from SS7. The 4-bit-long SSF is part of the MSUs, and its sole content is the network indicator (NI), which allows for national messages to be distinguished from international ones. The SSF is also part of the service information octet (SIO). See Chapter 8.

SSN [Subsystem Number Q.711 through Q.714, Send Sequence Number GSM 04.08] Subsystem number. Identifies the actual user of an SCCP message. Examples of SSNs are the BSSAP (see *BSSMAP*) and MAP. See Chapter 9. Also abbreviation for send sequence number (SSN), which relates to bit number 6 of the octet that indicates the message type of GSM messages on the Air-interface.

Stealing flag [GSM 05.03] The stealing flag is 1 bit long and is part of the normal burst. Stealing flags embrace the training sequence (see *TSC*). They are used in the uplink and downlink direction to indicate whether and which bits of a traffic channel are used (stolen) to carry signaling information (see *FACCH*). Stealing of bits to send signaling information on the traffic channel (see TCH) may become necessary when the MS or the BTS have to immediately send signaling data but the SACCH is not available.

Examples of control data are the HND_CMD, CON, and the DISC messages. Note that 456 bits (four bursts) are necessary to transfer such a message, completely. The stealing flag was introduced in order not to lose four consecutive bursts for traffic data. Actually, a signaling message is divided into eight packets with 57 bits each and then transmitted in eight consecutive bursts. Only the even-numbered bits are used to carry signaling data in the first four bursts, while only the odd-numbered bits are used for signaling in the last four bursts (i.e., 5 through 8). That allows use of the remaining 57 bits per burst to carry traffic data (Figure G.55).

STP Signaling transfer point. A node within a SS7 network. The STP does not only process SS7 messages, it also relays them to other SS7 network elements (Figure G.56). Basically, every SS7 network element could be used as an STP, too. See Chapter 8.

Subchannel [GSM 05.01, 05.02] For the SDCCH and for the halfrate traffic channel, a single Air-interface channel with 13 Kbps can be divided into subchannels. In halfrate, one channel simply is divided in half, that is, two subchannels, while in SDCCH up to eight subchannels can be configured on a single 13-Kbps channel. See Chapter 7.

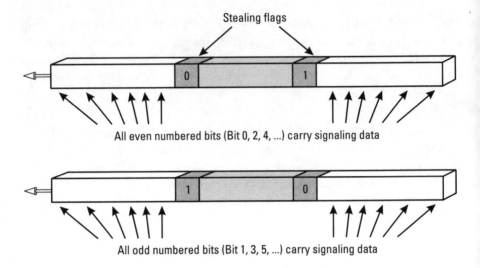

Figure G.55 The stealing flag.

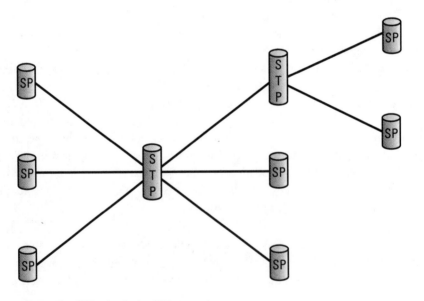

Figure G.56 The STP relay in the SS7 network.

Subscriber An entity that subscribes to telecommunications services. A subscriber can be a natural person or a legal entity. GSM does not distinguish between the subscriber and the user of a service.

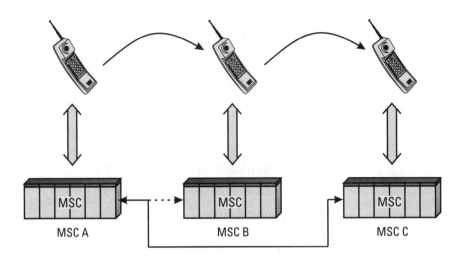

Figure G.57 Intersystem and subsequent handover.

Subsequent handover [GSM 09.10] Situation in which an inter-MSC handover occurs more than once for a single connection. MSC A hands the call to MSC B, later the call is handed over to MSC C or back to MSC A (Figure G.57). Since MSC A is always in charge of such a connection, the scenario for the second or third handover is different from the first one. For that reason, the term *subsequent handover* was introduced. See Chapter 12.

SUERM [Q.700–Q.704] Signal unit error rate monitor. Part of Layer 2 of an SS7 network element. The number of faulty signal units received subsequently or within a certain time period is monitored. When a defined threshold is passed, for example, 64 subsequent faulty signal units have been received, an error indication is issued to Layer 3, and the SS7 connection is taken out of service.

Superframe [GSM 05.01, 05.02] A superframe consists of twenty-six 51-multiframes or fifty-one 26-multiframes and has a length of 6.12s. See Chapter 7.

Supplementary services See *SS*.

Synchronization channel See *SCH*.

Synchronized handover [GSM 04.08] All BTSs that are finely synchronized can perform a so-called synchronized handover. Fine synchronization means that the starting times and the ending times of all time slots are identical for the synchronized cells. That is not the case when the two BTSs are not finely synchronized. Compare in Figure G.58 the channel structure of BTS 1, BTS 2, and BTS 3. While BTS 1 and BTS 3 are finely synchronized, BTS 2 operates independently.

Now let's assume that a handover is to be performed from BTS 1 to BTS 3. In general, the distance from the MS to BTS 1 and to BTS 3 will be different. Of course, that results in different propagation delays of the signal from BTS 1 to the MS, compared to the time the signal takes from BTS 3 to the MS. Based on that difference, the MS is able to calculate the distance to BTS 3 and to adjust its timing advance (see *TA*) for a handover from BTS 1 to BTS 3. Consequently, it is not necessary for BTS 3 to send PHYS_INFO messages to the mobile station.

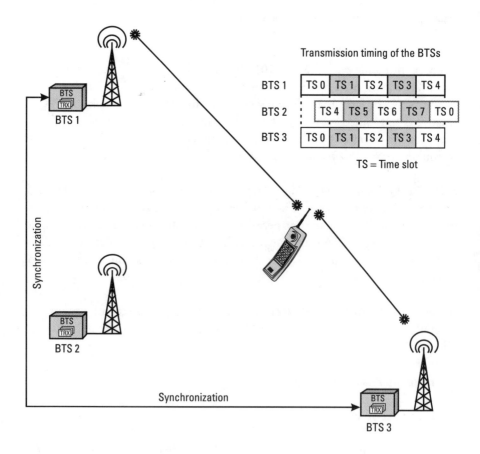

Figure G.58 Synchronized BTSs.

This possibility does not exist for a handover from BTS 1/BTS 3 to the nonsynchronized BTS 2. Only a nonsynchronized handover is possible in this case and requires BTS 2 to send PHYS_INFO messages during the handover, for the MS to adjust the TA.

GSM has defined two more handover types, the pseudo-synchronized and the presynchronized handover. The pseudo-synchronized handover requires the MS to predetermine the time offset between BTS 1 (active cell) and BTS 2 (destination cell) from the neighbor cell measurements. The MS sends the resulting MEAS_REP messages to BTS 1 or, more precisely, to the BSC. These data reflect the view of the MS, only. Another condition for pseudo-synchronized handover is that BTS 1 needs to know exactly how much earlier or later the destination cell (BTS 2), relative to BTS 1, actually transmits. With that information and with the results of the measurements from the MS it is possible for BTS 1 to calculate the distance between MS and BTS 2 and, hence, the TA that the MS has to preset when being handed over to BTS 2. BTS 1 sends the TA value to be used in an HND_CMD message to the MS. For the MS, pseudo-synchronized handover is an optional feature, and its availability is indicated to the BSC in the mobile station classmark during call setup.

A presynchronized handover requires the same behavior from the MS as a synchronized handover. The BTS does not send a PHYS_INFO message and the MS uses either the value requested in the HND_CMD message or a standard TA, with a value of 1, when the HND_CMD message does not contain any value for TA.

SYSTEM Information 1–8 See BCCH SYS_INFOS 1–4.

T7 [GSM 08.08] Timer T7 is a timer of the BSC that indicates the minimum time interval between two HND_RQD messages for the same connection (Figure G.59).

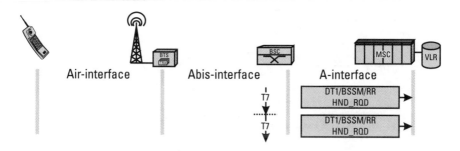

Figure G.59 Timer T7 in the BSC.

T8 [GSM 08.08] A timer of the "old" BSC that indicates the waiting period during an external handover between transmission of the HND_CMD message and either the positive acknowledgment of a handover by the MSC (CLR_CMD) or the negative acknowledgment by the MS (HND_FAI). The old BSC releases the connection with a CLR_REQ message (cause 0 = radio interface message failure) in case T8 expires, before any of these messages is received. See Figure G.60.

T10 [GSM 08.08] A timer of the BSC that supervises the waiting period between emission of an ASS_CMD message and the according acknowledgment by the MS (Figure G.61). The BSC sends an ASS_FAIL cause: radio interface message failure to the MSC, if the timer expires during a TCH assignment, without an ASS_COM message or an ASS_FAI message being received from the MS.

If, during intra-BTS handover, an HND_CMD message instead of the ASS_CMD message is used for the assignment of the new traffic channel, the BSC sends CLR_REQ cause: radio interface message failure to the MSC, instead of the ASS_FAIL message.

T200 [GSM 04.06, Q.921] Timer used by LAPD and $LAPD_m$. The sender of a frame expects to receive acknowledgment for that frame within the time specified by T200. T200 starts when a message is sent. If no acknowledgement is received within T200, the frame may be retransmitted as many times as specified by N200, before the connection is released.

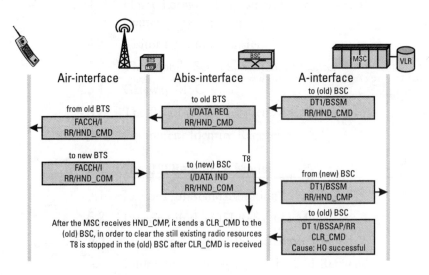

Figure G.60 The task of timer T8.

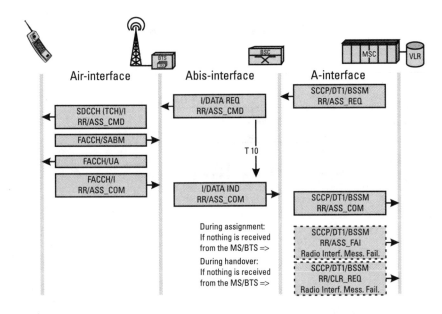

Figure G.61 Timer T10 during the TCH assignment.

T3105 [GSM 04.08] A timer of the BTS that defines the time between transmission of two consecutive PHYS_INFO messages.

T3212 [GSM 04.08] A timer of the mobile station or the SIM. The value of T3212 specifies the time interval for periodic location updating.

TA Timing advance. The agreement in a GSM system is for the MS to send its data three time slots after it received the data from the BTS. The BTS then expects the bursts from the MS in a well-defined time frame. This prevents collision with data from other mobile stations. The mechanism works fine, as long as the distance between MS and BTS is rather small. Increasing distance requires taking into account the propagation delay of downlink bursts and uplink bursts. Consequently, the mobile station needs to transmit earlier than defined by the "three time slots delay" rule. The information about how much earlier a burst has to be sent is conveyed to the mobile station by the TA. The TA is dynamic and changes in time. Its current value is sent to the mobile station within the layer 1 header of each SACCH. In the opposite direction, the BTS sends the current value for TA within the MEAS_RES messages to the BSC (e.g., for handover consideration). The farther the MS is away from the BTS, the larger is the required TA. Figure G.62 illustrates the relation between distance and TA.

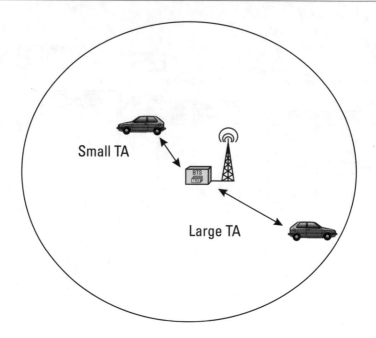

Figure G.62 The dependency of TA from the distance.

Using the TA allows the BTS to receive the bursts from a particular MS in the proper receiver window. The BTS calculates the first TA when receiving a RACH and reports the value to the BSC. TA can take any value between 0 and 63, which relates to a distance between 0 km and 35 km. The steps are about 550 m (35 km/63 ≈ 550 m). With respect to time, the different values of TA refer to the interval 0 μs through 232 μs, in steps of 48/13 μs. It is important to note that this value of TA represents twice the propagation delay. Figure G.63 illustrates the effect of TA by an example in which a connection is active on TS 1.

TAF [GSM 07.01, 07.02, 07.03] Terminal adaptation function. Function that supports the mobile station in setting up data connections.

TAG [Q.773] Indicates parameter types within TCAP and MAP. The term stems from ASN.1. See Chapter 11.

Tail bits See *Burst*.

TCAP [Q.771, Q.772, Q.773] Transaction capabilities application part. See Chapter 11.

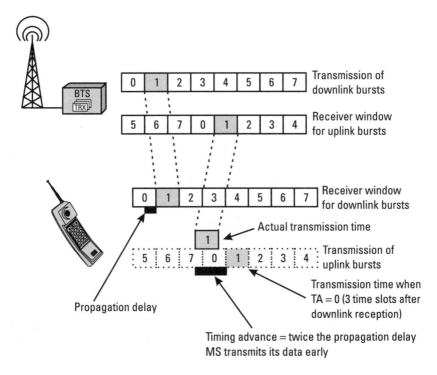

Figure G.63 Adjusted transmission time due to TA.

TCH [GSM 05.01, 05.02, 05.03] Traffic channel. GSM uses the following types of TCHs (see Table G.18).

Table G.18
The TCHs in GSM

Abbreviation	Remark
TCH/FS	Fullrate speech traffic channel
TCH/HS	Halfrate speech traffic channel
TCH/F9.6	9.6-Kbps fullrate data traffic channel
TCH/F4.8	4.8-Kbps fullrate data traffic channel
TCH/H4.8	4.8-Kbps halfrate data traffic channel
TCH/F2.4	2.4-Kbps fullrate data traffic channel
TCH/H2.4	2.4-Kbps halfrate data traffic channel

TDMA   Time division multiple access. See *FDMA*.

TEI [Q.920, Q.921]   Terminal endpoint identifier. Used by the LAPD protocol in connection with the SAPI to specify the destination of a message. See Chapter 6.

Teleservice   A telecommunications service that a GSM operator offers to its customers (subscribers). While for a teleservice all seven OSI layers are specified, the related bearer services require only Layers 1 through 3. A bearer service forms, from a physical point of view, the lower layers of a teleservice. From a standardization point of view, they belong, however, to different categories. Table G.19 lists the Teleservices of GSM.

TI   Transaction identifier.

Timing advance   See *TA*.

TMSI [GSM 03.03]   Temporary mobile subscriber identity. Used to identify a mobile subscriber, like the IMSI. Unlike the IMSI, however, the 4-byte-long TMSI has only temporary significance. The VLR assigns a TMSI upon location registration for confidentiality purposes, so it is not required to transfer the IMSI over the Air-interface frequently. Assignment and use of the TMSI is only possible with active ciphering. The TMSI can take any value, except FF FF FF FF$_{hex}$. This value is reserved in case the SIM does not contain a valid IMSI.

TN   Time slot number.

Table G.19
Teleservices in GSM

Teleservice	Code	Description
Speech	11	Regular telephone service
	12	Emergency calls
Short Message Services	21	SMS/(MT/PP)
	22	SMS/(MO/PP)
	23	Cell broadcast (CB)
Fax	61	Speech + fax/group 3
	62	Fax/group 3

Training sequence code See *TSC*.

Transaction identifier [GSM 04.07, 04.08] Parameter used by CC (call control) Layer 3 messages. Note that RR messages and MM messages do not need a transaction identifier and therefore fill its position with the skip indicator (always 0000_{bin}).

The purpose of the TI is to distinguish among simultaneous connections (transactions) of one mobile station. See Chapter 7.

TRAU Transcoding rate and adaptation unit. See Chapter 3

TRAU frame [GSM 08.60] In fullrate channels, every TRAU frame is 320 bits long. In halfrate channels, both 160 bits and 320 bits are possible. One has to distinguish four different types of TRAU frames:

- The speech frame is used to transfer voice. The payload is 260 bits for fullrate and 112 bits for halfrate.
- The O&M frame to transfer O&M data between BTS and TRAU. The BTS acts in this case only as a relay for the BSC, for which these data are actually destined or where they are generated (264-bit O&M data for both fullrate and halfrate).
- The data frame carries 252 bits of payload in fullrate and 126 bits in halfrate, but no voice.
- The idle speech frame for quiet times between BTS and TRAU with active DTX.

TS Time slot.

TSC [GSM 05.02] Training sequence code. Code positioned in the middle of every burst, independent from the direction uplink or downlink. The dummy burst and the frequency correction burst are both exceptions from this rule and do not contain a TSC. The length of the TSC depends on the burst type. The TSC can be of fixed format or be assigned (there are eight TSC formats) (Figure G.64).

- Synchronization burst (see *Burst*): predefined TSC with 64 bits.
- Access burst: predefined TSC (synchronization sequence) with 41 bits.
- Normal burst: dynamic TSC with 26 bits; can be independently assigned for common control channels (see *CCCH*) and dedicated

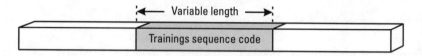

Figure G.64 The TSC.

channels (see *TCH*, *SDCCH*). Which of the eight TSCs is used for CCCHs, is determined by the base station color code (see *BCC*). Assignment of the TSC for dedicated channels takes place in the IMM_ASS message, the ASS_CMD message, and the HND_CMD message.

Since the content of the TSC is known on both sides, it can be applied as a reference that enables the receiver to conclude on possible bit errors of the entire burst. Depending on the grade and type of error, either the original data can be restored or the whole burst will be discarded. The bit error rate that was determined by means of the TSC is the basis for the RXQUAL information of the MEAS_RES messages and the MEAS_REP messages.

Triplet Generally refers to three related things. Used in GSM in relation to security features, where it refers to SRES, RAND, and K_c.

TRX Transmission/reception unit. Generic term for both the high-frequency and low-frequency modules that perform signal processing in the BTS. Some of the tasks of the TRX are channel coding/decoding, ciphering/deciphering, and GMSK modulation/demodulation. See Chapter 3.

TX_Integer [GSM 05.08] Parameter transmitted in the RACH control parameters in the BCCH / SYS_INFO 1–4. TX_Integer indicates the maximum number of RACH time slots that a mobile station has to wait after an unsuccessful BTS access before another CHAN_REQ can be sent. See Chapter 7.

In those cases, the MS selects a random number R, with

$$0 \leq R \leq TX\_Integer$$

and waits R TDMA frames before the CHAN_REQ messages is sent again (see MAX_RETRAN).

μ-law   See *A-law*.

Um-interface   The GSM-specific term for the Air-interface.

UMTS   Universal Mobile Telecommunication System (UMTS). A third-generation mobile telecommunication system currently being discussed by SMG as an evolutionary step of GSM. The goals of UMTS are similar to those of the Future Public Land Mobile Telecommunication System (FPLMTS).

Uplink   The direction of a signal from the mobile station toward the network. See *Downlink*.

User parts   The various applications of the MTP (SS7). Examples are the SCCP and ISUP. See Chapters 8 and 9.

USSD [GSM 02.90]   Unstructured supplementary service data (USSD). In addition to the supplementary services, which are defined by GSM 02.04, GSM has defined the concept of the unstructured supplementary services (USSs). USSs allow the network operators to offer PLMN-specific supplementary services to their customers with a rather short time to market. USSs can be implemented in the MSC, the VLR, or the HLR. For the USSs, some MAP messages have been predefined that allow data packages of up to 80 bytes to be sent. The data packages are referred to as USSD. The subscriber is capable of entering USSD at the mobile station via the HMI.

VAD [GSM 06.32]   Voice activity detector. A unit in the BTS and the MS that determines whether or not voice activity prevails. If there is no voice signal, it will be switched to DTX mode.

VLR [GSM 03.02, 12.02, 12.03]   Visitor location register. See Chapter 4.

VLR number [GSM 03.03]   An identifier for the VLR that corresponds to that of a regular ISDN number or directory number (see *MSISDN*). Every VLR has a unique VLR number that is used to address messages to the VLR via the SCCP.

About the Author

Gunnar Heine received his Ph.D. in Communications Technology from the University of Wilhelmshaven. In 1991, he began his career as a system engineer of GSM for ALCATEL SEL in Germany. Heine was responsible for the integration and troubleshooting of the just commissioned GSM systems both in Germany and internationally. In 1996, Gunnar Heine was sent to the United States to take over the position of Director of Technical Support for ALCATEL Mobile Switching. After 7 1/2 years with ALCATEL, he has returned to Germany and is currently the VP of Technology Management for HanseNet GmbH in Hamburg.

Index

A-format frame, 102, 105
A-interface, 27, 168, 277, 280
    base station subsystem application
        part, 173–84
    dimensioning, 171–73
    error analysis, 287–91
    error analysis in BSS, 296–302
    error messages, 291–96
    signaling analysis, 280–82, 285
Abis interface, 19, 25, 26–27, 44, 277, 280
    BTS/BSC connection, 52–55
    channel configurations, 51–52
    error analysis, 287–91
    error analysis in BSS, 296–302
    error messages, 291–96
    layer 2, 56–71, 81–83
    layer 3, 71–86
    protocol stack, 55–56
    service, bringing into, 87
    signaling analysis, 280, 282–84, 286
ABORT message, 221
Abstract Syntax Notation
    number 1, 45, 189, 193, 194
Access grant channel, 95–100
ACCH. *See* Associated control channel
Acknowledged message, 57
Address field, 60, 83, 102, 104, 105
Addressing
    Signaling System Number 7, 130–33

transaction capabilities application
    part, 186–88
AGCH. *See* Access grant channel
Air-interface, 19, 27, 28, 44, 63, 71, 75, 279
    channel configuration, 97–100
    interleaving, 100–1
    layer 2 signaling, 101–6
    layer 3 signaling, 107–23
    logical-channel connection, 94–100
    physical vs. logical channel, 94
    structure, 89–94
ALCATEL, 34
AM. *See* Amplitude modulation
Amplitude modulation, 90
APDU. *See* Application protocol data unit
Application layer, 46
Application protocol data unit, 189
Architecture, 3–4
ASN.1. *See* Abstract Syntax Notation
    number 1
Assignment failure
    message, 292–94, 298–300
Associated control channel, 102
AuC. *See* Authentication center
Authentication center, 32–33

B-format frame, 102, 105
B-interface, 34, 37, 51, 214
Backward indicator bit, 127, 135–38

Backward sequence number, 127, 133–38
Base station controller, 5–6, 25–27,
  61, 73, 130, 172–73, 227
  connection to BTS, 52–55, 87
  handover, 257, 260–62
Base station range, 3
Base station subsystem, 6, 29–30, 174
  error analysis, 280, 287, 296–302
  location update, 227–31
  mobile originating call in, 233, 235–40
  mobile terminating call in, 244–50
Base station subsystem application
  part, 153–54, 169,
  173–76, 285
Base station subsystem management
  application part, 173–75
  message decoding, 183–184
  message types, 176–183
Base transceiver station, 5, 17, 19–25,
  61, 64, 72–73, 75, 91, 171,
  227, 287
  clock module, 21–22
  configurations, 22–25
  connection to BSC, 52–55, 87
  handover, 251, 254, 257–59
  input/output filters, 22
  operations and maintenance, 20
  transmitter/receiver, 20
Bbis-format frame, 102
BCCH. *See* Broadcast common
  control channel
Begin message, 201, 221
BIB. *See* Backward indicator bit
Binary data, 285
Bit, 9
Bit error rate, 296
Black list, 38
Broadcast common control
  channel, 95–100
BSC. *See* Base station controller
BSN. *See* Backward sequence number
BSS. *See* Base station subsystem
BSSAP. *See* Base station subsystem
  application part
BSSMAP. *See* Base station subsystem
  management application part
BTS. *See* Base transceiver station

Burst, 90–91
Burst segmentation, 105
Call control, 7, 17, 44, 104, 107–10,
  118–23, 174, 263, 267
Called-party address, 159, 162–65
Call handling error, 21–22
Calling-party address, 159, 162–65
Call release, 10
Call setup, 10
CaPA. *See* Calling-party address
CBA. *See* Change back acknowledge
CBCH. *See* Cell broadcast channel
CBD. *See* Change back declaration
CC. *See* Call control; Connection confirm
CCCH. *See* Common control channel
CCH. *See* Control channel
CCM. *See* Common channel management
CDMA. *See* Code division multiple access
CdPA. *See* Called party address
Cell broadcast channel, 95–100
Cell identity, 22
Cellular network, 3–4
Central module, base station controller, 27
CEPT. *See* Conference Européenne des
  Postes et Télécommunications
Change back acknowledge, 148
Change back declaration, 148
Change over acknowledge, 148
Change over order, 140, 148
Changeover procedure, 140–41
Channel coding, 105
Channel mapping, 91
Channel number, 75
Checksum, 44
CI. *See* Cell identity
Ciphering, 91
Class field, 191
Clear request message, 291–92, 297–98
Clear-to-send signal, 43
Clock module, 21–22, 91
CNP. *See* Connection not possible message
CNS. *See* Connection not successful message
COA. *See* change over acknowledge
Code division multiple access, 2, 89
Coding, transaction capabilities application
  part, 189–98

Collocated base transceiver station, 24–25, 54
Command/response bit, 61, 62, 102, 106
Common channel management, 55, 75
Common control channel, 75
Component portion, 205–8
Compression. *See* Speech compression/decompression
Computer port, 43
Conference Européenne des Postes et Télécommunications, 2
Connect message, 201
Connection confirm, 161, 167
Connection failure message, 294–95
Connectionless service, 154–56, 165–66, 186, 285, 287
Connection non successful message, 149
Connection not possible message, 149
Connection-oriented service, 154–56, 165–66, 285–86
Connection refused message, 161
Connection request message, 161, 167
Connection successful message, 149
Consistency check, 57, 64
Constructor, 191–93, 197
Control channel, 10, 94, 286
Control field, 62, 67, 69, 83
COO. *See* Change over order
CR. *See* Connection request
C/R bit. *See* Command/response bit
Credit parameter, 165
CREF. *See* Connection refused message
CSS. *See* Connection successful message
CTS signal. *See* Clear-to-send signal

D-channel, 10, 55–57
 See also Data Link Access for D-Channel
Database, base station controller, 27
Data form 1, 161, 168
Data link connection, 149
Data link connection identifier, 175
Data link layer, 43–44
Data type octetstring, 194
DCCH. *See* Dedicated control channel
DCM. *See* Dedicated channel management
DCS 1800. *See* Digital Cellular System 1800

DECT. *See* Digital enhanced cordless telecommunications
Dedicated channel management, 55, 75
Dedicated control channel, 102
Destination local reference, 154–56, 166–68, 189, 285
Destination point code, 131–32
DeTeMobil software, 280–84
Dialog control, 198–99
 structured vs. unstructured, 198, 201, 220–23
Dialog portion, 198, 203–4
Dialog unit, 201–2
Digital Cellular System 1800, 2
Digital Enhanced Cordless Telecommunications, 2
Dimensioning, 171, 173
Direction dependency, 211–12
Direct transfer application part, 173–75
Disconnected-mode frame, 56, 68
Disconnect frame, 56, 69
Discontinuous transmission, 29, 255–56
DLC. *See* Data link connection
DLCI. *See* Data link connection identifier
DLR. *See* Destination local reference
DM frame. *See* Disconnected–mode frame
Downlink, 73, 94–95, 97–98, 102
 synchronization, 93–94
DPC. *See* Destination point code
Drivetest, 275–76
DT 1. *See* Data form 1
DTAP. *See* Direct transfer application part
DTMF. *See* Dual-tone multifrequency
DTX. *See* Discontinuous transmission
Dual-tone multifrequency, 17
Duplex service, 90

EA field-bit. *See* Extension address field-bit
ECA. *See* Emergency change over acknowledge
ECO. *See* Emergency change over order
EIR. *See* Equipment identity register
EL-bit, 104
Emergency change over order, 148
End message, 208, 210, 221
End-of-optional-parameter, 158, 165
End-to-end addressing, 163

End-to-end data control, 44–45
EO. *See* End-of-optional-parameter
Equipment check, 227, 233
Equipment identity
    register, 7, 31, 37–38, 185,
    227, 233
Error analysis
  in base station subsystem, 296–302
  general information, 287–91
  important messages, 291–96
Error detection/correction, 43–44, 133–38
Error indication message, 296
ETSI. *See* European Telecommunications Standard Institute
European Telecommunications Standard Institute, 2
Exchange-identification frame, 57, 70
Extension address field-bit, 61, 102

F-bit. *See* Final bit
FACCH. *See* Fast associated control channel
Fast associated control channel, 95–100
FCCH. *See* Frequency correction channel
FCS. *See* Frame check sequence
FDMA. *See* Frame division multiple access
FIB. *See* Forward indicator bit
Fifty-one multi-frame, 91–92, 97–99
Fill-in octet, 105–6
Fill-in signal unit, 127, 130, 134, 137–38
Final bit, 62
Fine synchronization, 22
FISU. *See* Fill-in signal unit
Fixed mapping, 51–52
Flag parameter, 60, 105
Format field, 191
ForwardAccessSignaling service, 213
Forward indicator bit, 127, 135–38
Forward sequence number, 127, 133–38
Frame check sequence, 43–44, 60, 63,
    82, 102, 105, 127, 133, 137
Frame hierarchy, 90–93
Frame length indicator, 104
Frame-reject frame, 57, 69–70
Frequency allocation, 3
Frequency correction channel, 95–100
Frequency division multiple access, 89–91
Frequency hopping, 20–21

FRMR. *See* Frame-reject frame
FSN. *See* Forward sequence number
Fullrate channel, 28–29, 90

Gateway mobile switching center, 36, 251
Gaussian minimum shift
    keying, 17, 20, 43, 90
Global System for Mobile
    Communication, 1–3
  architecture, 3–4
  subsystem overview, 4–7
Global title, 162–63
GMSK. *See* Gaussian minimum shift keying
Gray list, 38
GSM. *See* Global System for Mobile Communication

H-interface, 32
Halfrate channel, 28–29
Handoff, 251
Handover, 21, 24, 27, 36
  analysis, 255–56, 294
  external, 260
  inter-MSC, 263, 267–69
  intra-BSC, 257, 260–62
  intra-BTS, 257–59
  intra-MSC, 260, 263–66
  measurement of BTS/MS, 251, 254
  subsequent, 267, 270–72
  synchronized/nonsynchronized, 22, 24,
    91–93, 256–57, 261, 263
HDLC. *See* High-level data link control
Heading code, 142
Hexadecimal data, 285
HF. *See* High frequency
High frequency, 43
High-level data link control, 43, 56
HLR. *See* Home location register
HMI. *See* Human-machine interface
Home location register, 5–6, 31–35,
    130, 185, 202, 205, 221, 244
Human-machine interface, 20, 86
Hyperframe, 91, 93

I-frame, 66–67, 71
Idle frame number, 97
Idle phase, 64, 67
IEI. *See* Information element identifier

IMEI. *See* International mobile equipment identity
IMSI. *See* International mobile subscriber identity
Inactivity test, 162
Inband signaling, 10, 29
Information element identifier, 175
Information field, 105
Input filter, 22
Integrated Services Digital Network, 34, 56, 108
Interfaces, network switching subsystem, 31–32
Interference, 3, 53, 279
Interleaving, 100–1, 105
International mobile equipment identity, 7, 18, 37–38, 227, 234
International mobile subscriber identity, 191–92, 202, 205, 227
International Telecommunications Union, 1, 45, 46, 51, 56, 61, 63, 139, 153, 158, 164, 186, 189, 194
International Telecommunications Union Telecommunication Standardization Sector, 10, 198
Interworking function, 36, 40
Invoke component, 205
IS-95 standard, 89
ISDN. *See* Integrated Services Digital Network
ISDN user part, 36, 125, 128, 130, 153, 233, 251
ISUP. *See* ISDN user part
IT. *See* Inactivity test
ITU. *See* International Telecommunications Union
ITU-T. *See* International Telecommunications Union Telecommunication Standardization Sector
IWF. *See* Interworking function

Japan, 3

LAPD. *See* Link Access Protocol for D-channel

"Leap," layer, 47
Least significant bit, 194
Length indicator, 196–98
LFU. *See* Link forced uninhibit
LIA. *See* Link inhibit acknowledgment
LID. *See* Link inhibit denied
LIN. *See* Link inhibit signal
Line topology, 54
Link Access Protocol for D-channel, 10, 26, 43–44, 55–57, 211
    frame concept, 56–57
    frame types, 64–70
    layer 2 signaling, 101–6
    layer 3 signaling, 107–23
    message decoding, 70–71
    message parameters, 60–70
    modulo 128 vs. modulo 8, 57–59, 63–64
    vs. LAPDm, 105–6
Link Access Protocol for D-Channel modified, 101–6, 296
Link failure, 141–42
Link forced uninhibit, 150
Link inhibit acknowledgment, 150
Link inhibit denied, 150
Link inhibit signal, 150
Link local inhibit test, 150
Link remote inhibit test, 151
Link status signal unit, 127–28, 130, 132, 134
Link uninhibit acknowledgment, 150
Link uninhibit signal, 150
LLT. *See* Link local inhibit test
LMSI. *See* Local mobile subscriber identity
Local error code, 205, 207
Local mobile subscriber identity, 221–22
Local operation code, 213–20
Location update, 45, 227–32
Logical channel
    configuration, 94–100
    vs. physical channel, 94
LRT. *See* Link remote inhibit test
LSB. *See* Least significant bit
LSSU. *See* Link status signal unit
LU. *See* Location update
LUA. *See* Link uninhibit acknowledgment
LUN. *See* Link uninhibit signal

M-bit, 104
Management message, 142–51
Mandatory fixed part, 158
Mandatory variable part, 158
MAP. *See* Mobile application part
MAP-CLOSE service, 213
MAP-DELIMITER service, 212
MAP-NOTICE service, 213
MAP-OPEN service, 212
Mapping, logical channel, 95–100
MAP-U-ABORT service, 213
ME. *See* Mobile equipment
MEAS_REP analysis, 256
MEAS_RES analysis, 255
Message discriminator, 73, 75
Message representation, 10–12
Message signal unit, 127–30, 134–38, 156
Message transfer, 135–38
Message transfer part, 44, 126–27, 153
Message types
    air interface, 109–13
    base station subsystem management
        application part, 176–84
    connection principle, 167–69
    radio link management, 75–80
    signaling connection control
        part, 161–67
    Signaling System
        Number 7, 127–30, 148–51
    transaction capabilities application
        part, 198–200
Microprocessor, 39
MM. *See* Mobility management
Mobile application
    part, 153, 163–65, 185, 205,
        208–9, 285
    application communication, 220–23
    local operation codes, 214–20
    services, 211–14
    user communication, 209–11
Mobile application part application, 211
Mobile equipment, 227
Mobile originating call, 45, 108, 233–43
Mobile services switching center, 6
Mobile station, 5, 14, 17–18, 75, 109
    handover measurement, 251, 254
Mobile station roaming number, 244

Mobile subscriber identification
    number, 192, 193
Mobile subscriber ISDN, 244
Mobile switching center, 6, 31, 34–37,
    75, 130, 137–38, 169, 171–73,
    185, 227
    gateway, 36, 251
    handover, 260, 263–69
    visitor location register, 36–37
Mobile terminating call, 45, 244–53
Mobility management, 7, 17, 44,
    104, 108–18, 174
MOC. *See* Mobile originating call
Modularization, 39
Most significant bit, 194
MS. *See* Mobile station
MSB. *See* Most significant bit
MSC. *See* Mobile switching center
MSIN. *See* Mobile subscriber
    identification number
MSISDN. *See* Mobile subscriber ISDN
MSRN. *See* Mobile station roaming number
MSU. *See* Message signal unit
MT. *See* Message type
MTC. *See* Mobile terminating call
MTP. *See* Message transfer part
Multiframe, 91, 97–99
MYC. *See* Mobile terminating call

National destination code, 244
NDC. *See* National destination code
Network indicator, 130
Network layer, 44
Network management, Signaling System
    Number 7, 138–51
Network switching
    subsystem, 31–32, 109, 153,
        185, 285
    location update, 227, 232–33
    mobile originating call in, 233, 242–43
    mobile terminating call in, 244, 251–53
    signaling analysis, 280
NI. *See* Network indicator
N(R). *See* Receive sequence number
N(S). *See* Send sequence number
NSS. *See* Network switching subsystem

O&M module. *See* Operations and
  maintenance module
OACSU. *See* Off-air call setup
Octetstring, 194
Off-air call setup, 226
OMC. *See* Operations and
  maintenance center
OML. *See* Operation and maintenance link
OPC. *See* Originating point code
Open System Interconnection Reference
  Model, 7, 39–40
  advantages, 42–43
  comprehension issues, 46–49
  data types, 41
  information processing, 42
  layer 2, Abis-interface, 56–71
  layer 3, Abis-interface, 71–86
  layered architecture, 40–41, 43–46
  protocol stack, 55–56
Operation and maintenance link, 83–85
Operations and maintenance
  center, 22, 27, 72, 276–78
Operations and maintenance link, 83–86
Operations and maintenance
  module, 20, 83–86
Optional part, 158
Originating point code, 130–32
OSI Reference Model. *See* Open System
  Interconnection Reference Model
Outage, 140–41
Outband signaling, 10
Output filter, 22
Overload, 67, 89–90, 139–40

Paging channel, 95–100
P-bit. *See* Polling bit
PC. *See* Problem code
PCH. *See* Paging channel
PCM. *See* Pulse code modulation
PC parameter. *See* Protocol class parameter
PCS 1900. *See* Personal Communication
  System 1900
PD. *See* Protocol discriminator
Peer-to-peer protocol, 41, 209
Personal Communication System 1900, 2
Personal Handy Phone System, 2–3
P/F bit. *See* Polling/final bit

PHS. *See* Personal Handy Phone System
Physical channel, 94
Physical layer, 43
PLMN. *See* Public land mobile network
Point code, 130
Pointer, 158
Point-to-point connection, 105
Polling bit, 62–63
Polling/final bit, 62, 106
Polling process, 64
Power control input parameters, 254
Presentation layer, 45–46
Primitive, 191–93, 211, 221
Problem code, 208–9
Protocol analyzer, 276–79, 287
Protocol class parameter, 154, 165–66
Protocol discriminator, 107, 184
Protocol measurement, 44
  operations and maintenance
    center, 276–78
  protocol analyzer, 278–79
  tools, 275–76
Protocol stack, 55–56
PSTN. *See* Public switched telephone
  network
Public land mobile network, 18, 32
Public switched telephone network, 125
Pulse code modulation, 9–10, 43, 171
Pulse code modulation clock, 21

QOS. *See* Quality of service
Quality of service, 53, 100–1, 275–76

RACH. *See* Random access channel
Radio interface failure, 292
Radio interface message failure, 292
Radio link management, 55, 75
  message types, 75–80
Radio resource management, 108–10
Radio signaling link, 73–74, 84
RAND. *See* Random number
Random access channel, 95–100
Random number, 33
RC. *See* Release cause
RCT. *See* Route set congestion test
Ready-to-send signal, 43
Receive-not-ready frame, 56, 67–68

Receive-ready frame, 44, 56, 63–64,
    66–67, 86, 104, 263
Receive sequence number, 63, 67–68, 135
Refusal cause, 166
Regular pulse excitation-long term
    prediction, 28
Reject component, 208–9
Reject frame, 56, 67–68
Release cause, 166
Release complete, 161, 169
Released message, 161, 169
Return cause, 166
Return error component, 205
Return result component, 205, 208
RF. *See* Refusal cause
Ring topology, 54–55
RLC. *See* Release complete
RLM. *See* Radio link management
RLSD message. *See* Released message
RNR. *See* Receive-not-ready frame
Route set congestion test, 151
Route set test, 140–41, 149
Routing, message, 130–33
RPE-LTP. *See* Regular pulse excitation-
    long term prediction
RR frame. *See* Receive-ready frame
RR management. *See* Radio resource
    management
RSL. *See* Radio signaling link
RST. *See* Route set test
RT. *See* Return cause
RTS signal. *See* Ready-to-send signal

S-bit, 75
SABM frame. *See*
Set-asynchronous-balance-mode frame
SABME frame. *See* Set-asynchronous-
    balance-mode-extended frame
SACCH. *See* Slow associated control channel
SAPI. *See* Service access point identifier
SCCP. *See* Signaling connection control part
SCH. *See* Synchronization channel
SDCCH. *See* Standalone dedicated
    control channel
Sectorized base transceiver
    station, 24–25, 54
Segmentation, packet, 44

Segmenting/reassembling, 166
Send receive number, 104–5, 127
Send sequence number, 63, 67–68,
    104–5, 127, 135
Sequencing/segmenting, 166
Serial configuration, 52–55
Service access point
    identifier, 58, 60–61, 71–73,
    84, 102, 104, 175
Service indicator, 130
Service information octet, 128–29, 139, 156
Session layer, 45
Set-asynchronous-balance-mode
    frame, 46–47
Set-asynchronous-balance-mode-extended
    frame, 56, 68, 87, 105
Short-message service, 18, 71, 104, 107
SI. *See* Service indicator; Status indication
SIB, 128–29
SIE, 129, 132
SIEMENS, 34, 280
SIF. *See* Signaling information field
Signaling technology performance/
    uses, 8–10
Signaling analysis, 280
    automatic, 280–84
    manual, 284–85
Signaling connection control
    part, 27, 125–26, 128, 130,
    133, 173–74, 185–87, 189,
    196, 211
    connection identification, 285–87
    message format, 156–58
    message types, 158–67
    tasks, 153–56
Signaling information field, 130, 156
Signaling link selection, 130–31, 301
Signaling link test acknowledge, 151
Signaling link test
    acknowledgment, 132, 134
Signaling link test message, 132, 134, 151
Signaling point, 125–26, 128, 130, 139–41
Signaling point
    code, 130, 132, 159, 162,
    164, 285
Signaling System Number 7, 10, 27, 44,
    125–26, 173, 185–86, 251

error detection/correction, 133–38
message routing, 130–33
message transfer part, 126–27
message types, 127–30
network management, 138–51
Signaling transfer point, 125–26, 128, 130, 139–41, 187
Signed response, 33, 300–1
SIM. *See* Subscriber identity module
SIN, 129, 132
SIO. *See* Service information octet
SIOS, 128–29, 132
Skip indicator, 108
Slow associated control channel, 95–100, 256
SLR. *See* Source local reference
SLS. *See* Signaling link selection
SLTA. *See* Signaling link test acknowledgment
SLTM. *See* Signaling link test message
SMG. *See* Special Mobile Group
SMS. *See* Short-message service
Source local reference, 154–56, 166, 167, 189, 285
SP. *See* Signaling point
SPC. *See* Signaling point code
Special Mobile Group, 2–3
Speech compression/decompression, 28–29
S/R. *See* Segmenting/reassembling
SRES. *See* Signed response
SS. *See* Sequencing/segmenting; Supplementary services
SS7. *See* Signaling System Number 7
SSA. *See* Subsystem allowed
SSF. *See* Subservice field
SSN. *See* Subsystem number
SSP. *See* Subsystem prohibited
SST. *See* Subsystem status test
Standalone dedicated control channel, 95–100, 105, 251, 294
Star configuration, 52, 55
Status indication, 128
STP. *See* Signaling transfer point
Subscriber identity module, 5, 13–17
    ID-1, 13, 15
    plug-in, 13, 15
Subscriber identity module card, 300

Subservice field, 130
Subsystem allowed, 162
Subsystem number, 60, 159, 162, 164, 166–67
Subsystem overview, 4–7
Subsystem prohibited, 162
Subsystem status test, 162
Superframe, 91
Supervisory frame, 57, 63, 104
Supplementary services, 71, 107, 123
Switch matrix, 26
Synchronization channel, 91–94, 95–100
Synchronized handover, 22, 24, 91–93

T-bit, 73, 75
TA. *See* Timing advance
TAG field, 190–91, 193–95, 196–98
TCAP. *See* Transaction capabilities application part
TCE. *See* Terminal control element
TCH. *See* Traffic channel
TDMA. *See* Time division multiple access
TEI. *See* Terminal endpoint identifier
Telephone system applications, 40
Terminal control element, 26–27
Terminal endpoint identifier, 61, 102, 105
Testing, network, 138–51
Test message, 142–43, 147–51
Test mobile station, 18
TFA. *See* Transfer allowed
TFC. *See* Transfer controlled
TFP. *See* Transfer prohibited
TFR. *See* Transfer restricted
TI. *See* Transaction identifier
Time division multiple access, 89–91, 96
Time-division multiplexing, 9–10
Time slot, 90, 93, 100, 277
Timing advance, 24, 93–94
TRA. *See* Traffic restart allowed
Traffic channel, 94, 251, 286
Traffic restart allowed, 141, 151
Transaction capabilities application part, 45, 130, 153–54, 185–86, 285
    addressing, 186–87
    coding, 189–98
    GSM messages, 198–208

Transaction capabilities application
    part (continued)
    internal structure, 187–89
Transaction capability application
    part, 45, 130
Transaction identifier, 108–9, 184
Transcoding rate and adaptation
    unit, 6, 28–30, 171, 294
Transfer allowed, 141
Transfer controlled, 140, 150
Transfer prohibited, 140, 141, 148
Transfer restricted, 140, 149
Transmitter/receiver, 17, 19–21,
    52–53, 61, 286, 302
Transport layer, 44–45
TRAU. *See* Transcoding rate and
    adaptation unit
TRX. *See* Transmitter/receiver
TS. *See* Time slot
Twenty-six multi-frame, 91–92

UA frame. *See* Unnumbered
    acknowledgment frame
UDT message. *See* Unit data message
UDTS. *See* Unit data service
UI frame, 56, 63, 68–69, 71

Umbrella cell configuration, 22–24
UMTS. *See* Universal Mobile
    Telecommunication System
Unacknowledged message, 57, 58
Unit data message, 158, 161, 186, 189, 196
Unit data service, 161
Universal Mobile Telecommunication
    System, 3
Unnumbered-acknowledgment
    frame, 57, 69, 105
UpdateLocation service, 213–14
Uplink, 29, 73, 94–95, 97, 102, 109, 256
    synchronization, 93–94
UPU. *See* User part unavailable
Urban area, 53
User part unavailable, 151

Virtual connection, 156
Visitor location register, 5, 6, 31–37, 163,
    185, 202, 214, 221, 227, 244
VLR. *See* Visitor location register

White list, 38

X-bit, 75
XID. *See* Exchange-identification frame

Recent Titles in the Artech House Mobile Communications Series

John Walker, Series Editor

Advances in Mobile Information Systems, John Walker, editor

CDMA for Wireless Personal Communications, Ramjee Prasad

CDMA Mobile Radio Design, John B. Groe and Lawrence E. Larson

CDMA RF System Engineering, Samuel C. Yang

CDMA Systems Engineering Handbook, Jhong S. Lee and Leonard E. Miller

Cell Planning for Wireless Communications, Manuel F. Cátedra and Jesús Pérez-Arriaga

Cellular Communications: Worldwide Market Development, Garry A. Garrard

Cellular Mobile Systems Engineering, Saleh Faruque

The Complete Wireless Communications Professional: A Guide for Engineers and Managers, William Webb

GSM and Personal Communications Handbook, Siegmund M. Redl, Matthias K. Weber, and Malcolm W. Oliphant

GSM Networks: Protocols, Terminology, and Implementation, Gunnar Heine

GSM System Engineering, Asha Mehrotra

Handbook of Land-Mobile Radio System Coverage, Garry C. Hess

Handbook of Mobile Radio Networks, Sami Tabbane

High-Speed Wireless ATM and LANs, Benny Bing

An Introduction to GSM, Siegmund M. Redl, Matthias K. Weber, and Malcolm W. Oliphant

Introduction to Mobile Communications Engineering, José M. Hernando and F. Pérez-Fontán

Introduction to Radio Propagation for Fixed and Mobile Communications, John Doble

Introduction to Wireless Local Loop, William Webb

IS-136 TDMA Technology, Economics, and Services, Lawrence Harte, Adrian Smith, and Charles A. Jacobs

Mobile Communications in the U.S. and Europe: Regulation, Technology, and Markets, Michael Paetsch

Mobile Data Communications Systems, Peter Wong and David Britland

Mobile Telecommunications: Standards, Regulation, and Applications, Rudi Bekkers and Jan Smits

Personal Wireless Communication With DECT and PWT, John Phillips and Gerard Mac Namee

Practical Wireless Data Modem Design, Jonathon Y. C. Cheah

Radio Propagation in Cellular Networks, Nathan Blaunstein

RDS: The Radio Data System, Dietmar Kopitz and Bev Marks

Resource Allocation in Hierarchical Cellular Systems, Lauro Ortigoza-Guerrero and A. Hamid Aghvami

RF and Microwave Circuit Design for Wireless Communications, Lawrence E. Larson, editor

Signal Processing Applications in CDMA Communications, Hui Liu

Spread Spectrum CDMA Systems for Wireless Communications, Savo G. Glisic and Branka Vucetic

Understanding Cellular Radio, William Webb

Understanding Digital PCS: The TDMA Standard, Cameron Kelly Coursey

Understanding GPS: Principles and Applications, Elliott D. Kaplan, editor

Understanding WAP: Wireless Applications, Devices, and Services, Marcel van der Heijden and Marcus Taylor, editors

Universal Wireless Personal Communications, Ramjee Prasad

Wideband CDMA for Third Generation Mobile Communications,
Tero Ojanperä and Ramjee Prasad, editors

Wireless Communications in Developing Countries: Cellular and Satellite Systems, Rachael E. Schwartz

Wireless Technician's Handbook, Andrew Miceli

For further information on these and other Artech House titles, including previously considered out-of-print books now available through our In-Print-Forever® (IPF®) program, contact:

Artech House	Artech House
685 Canton Street	46 Gillingham Street
Norwood, MA 02062	London SW1V 1AH UK
Phone: 781-769-9750	Phone: +44 (0)20 7596-8750
Fax: 781-769-6334	Fax: +44 (0)20 7630-0166
e-mail: artech@artechhouse.com	e-mail: artech-uk@artechhouse.com

Find us on the World Wide Web at:
www.artechhouse.com

- DTMF capability;
- Short message service (SMS) capability;
- Availability of the ciphering algorithms A5/1 and A5/2;
- Display capability for short messages, dialed numbers, and available PLMN;
- Support of emergency calls, even without the SIM inserted;
- Burned-in IMEI.

2.2.5 Mobile Stations as Test Equipment

An MS is a useful test tool for the laboratory testing of a new network interface. Several manufacturers offer, for that purpose, a semistationary MS, which allows manipulation of specific system parameters to test the behavior of new software or hardware.

Besides those complex and expensive pieces of equipment used mainly in laboratories, a number of standard mobile telephones exist, which can easily be modified with additional packages to act as mobile test equipment. So-in-general they are composed of a personal computer and MS and offer functionality to monitor signaling between the network and the MS. Usually, it is possible to represent the test results in tabular or graphical form for easy analysis.

Despite these advantages, the test mobile stations seldom are used for protocol and error analysis because the results are not representative from a statistical point of view and can be gathered only with substantial effort and cost.

Nonetheless, special test mobiles are necessary tools for network operators, to monitor coverage and evaluate the behavior of the network as a customer would experience it.